MAKING SENSE

CULTURE AND ECONOMIC LIFE

EDITORS
Frederick Wherry
Jennifer C. Lena

EDITORIAL BOARD
Gabriel Abend
Michel Anteby
Nina Bandelj
Shyon Baumann
Katherine Chen
Nigel Dodd
Amir Goldberg
David Grazian
Wendy Griswold
Brayden King
Charles Kirschbaum
Omar Lizardo
Bill Maurer
Elizabeth Pontikes
Gabriel Rossman
Lyn Spillman
Klaus Weber
Christine Williams
Viviana Zelizer

MAKING SENSE

Markets from Stories in New
Breast Cancer Therapeutics

SOPHIE MÜTZEL

STANFORD UNIVERSITY PRESS
Stanford, California

STANFORD UNIVERSITY PRESS
Stanford, California

©2023 by Sophie Mützel. All rights reserved.

No part of this book may be reproduced or transmitted in any form or by any means, electronic or mechanical, including photocopying and recording, or in any information storage or retrieval system without the prior written permission of Stanford University Press.

Printed in the United States of America on acid-free, archival-quality paper

Library of Congress Cataloging-in-Publication Data available from the Library of Congress.

LCCN: 2022011744

ISBN (cloth): 9781503632554

ISBN (paper): 9781503634060

ISBN (digital): 9781503634077

Cover design: Susan Zucker

Cover image: Cancer cells | shutterstock

Typeset by Newgen in Minion Pro 11/14

*To Micha and Theo and Greti
for all your love and support*

Contents

Acknowledgments ix

	Introduction	1
1	Markets and Stories	18
2	Breast Cancer Therapies and Innovation	46
3	A Market of Expectations	58
4	Making Sense of the Market	73
5	Patterns in Meaning-Making: Categories over Time	89
	Conclusion: Markets from Stories	124

Appendix A: Research Design and Data 139

Appendix B: Technical Details on Formal Analyses 145

Notes 149

Bibliography 161

Index 201

Acknowledgments

This book has been in the making for a long time, during which I have received the support of many. Their ideas and criticisms have greatly improved this project; of course, I am responsible for all remaining shortcomings.

First and foremost, I thank Klaus Eder and my colleagues at the Humboldt-Universität Berlin, Institut für Sozialwissenschaften, for their sustained interest in and support of my perspective on markets and stories. I thank Michael Hutter and my colleagues at the Wissenschaftszentrum Berlin für Sozialforschung (WZB) for the time and support to develop a perspective on newness and how to cope with large amounts of data. The WZB also financially supported my stay at the Center for European Studies (CES), Harvard University, a time and space that was invaluable for sharpening my argument, obtaining data, and running analyses. At Harvard, my thanks go in particular to Michèle Lamont, the members of the Culture and Social Analysis Workshop, as well as to the Sociology of Valuation group, for spot-on discussions that brought the project forward. Thanks also go to my fellow scholars at the CES, in particular Nicolas Dodier, for asking intriguing questions. I thank the cancer researchers at the Dana-Farber Cancer Institute who were willing to answer my questions about developments in the field. I was also fortunate to be able to use the Harvard University Libraries during my stay at the CES, and for shorter periods of time, those of my alma mater Columbia University. I am grateful for all the librarians' help to my inquiries in locating literature. I thank Anna Kaiser, Julian Jürgenmeyer, Reini Joosten, and Alex Flückiger for providing research assistance for this project. For technical support at crucial steps, I thank Bine Schmidt and Sophian Sørensen.

Different parts of this research project have been presented at workshops and conferences. I am particularly thankful for the comments and insights of organizers and participants at several EGOS meetings, a Process Symposium, ASA, and DGS conferences.

Columbia University has been a formative place for me prior to moving back to Europe. Peter Bearman, David Stark, and Harrison White have fundamentally shaped my sociological thinking, each in his own way, and I am very grateful for their support. Independently, Peter and David asked crucial questions at the early stages of the project and helped to drive it forward.

Jens Beckert has been very supportive of this project from early on. I am extremely grateful for that, as well as for the years of exchange on economic sociology and the sociology of markets. Workshops organized at or with the help of the Max-Planck-Institut für Gesellschaftsforschung advanced my work. I owe special thanks to Steffen Mau, who agreed to support my *Habilitation* at the Humboldt-Universität zu Berlin, for which an earlier version of this book served as the required document. I thank the members of the *Habilitationskommission* for their time and support, in particular Ron Breiger, whose brilliance I have admired since my first days on this journey of sociology back at Cornell. I am deeply indebted to Ron's unwavering support and encouragement. Thank you!

The manuscript was predominantly written while at my current academic home, at the Universität Luzern, Soziologisches Seminar. I thank my colleagues and students for their continued support and patience while I was writing. I also gratefully acknowledge financial support for editorial work from the FaFoFö fund of the School of Cultural and Social Sciences at the University of Lucerne.

Huge thanks go to the community of the Neu Schönau writing group and, more recently, the writing academic group online for all their support. Our collective joy of writing propelled this project forward. As this manuscript moves into production, special thanks go to Gabriel Abend, Laura Portwood-Stacer, the anonymous reviewers of the manuscript, the editors Frederick Wherry and Jennifer C. Lena of the Stanford University Press series *Culture and Economic Life*, Shanon Fitzpatrick, Marcela Maxfield, Sunna Juhn, Emily Smith, Erin Ivy, as well as copy editors and graphic designers. It was a pleasure working with you.

My family and friends have also been incredibly supportive throughout the years. Thank you! Finally, I want to thank Micha, Theo, and Greti for their love, their endless support and patience (this is the book Mama was working on!), and all the happiness we have together.

MAKING SENSE

Introduction

BREAST CANCER IS ONE of the most commonly diagnosed cancers and a leading cause of death for women worldwide. Each year, more than 2 million new cases are diagnosed and more than 650,000 women die of breast cancer (Sung et al. 2021). Up until the 1970s, breast cancer could solely be treated with toxic and invasive remedies, i.e., surgery, radiation, and chemotherapy, in regimens that varied little from woman to woman. In the 1970s, hormonal therapies began to be developed to help limit the growth and spread of breast cancer cells. With discovery of the oncogene, a mutated gene that could lead to cancer, and advances of molecular engineering in the 1980s, hopes of patients and industry began to grow that a widely working, nontoxic, and noninvasive treatment for breast cancer could be developed. After twenty-five years of developments, for a limited percentage of breast cancer patients, six of such life-prolonging therapeutics had received regulatory approval, amounting to global sales of more than US$ 6.3 billion (Syed 2015).[1]

The search for such breakthrough therapeutics for breast cancer, like other quests for drug discovery and development, took many twists and turns. The many human and nonhuman actors who contributed to the development of a market for innovative breast cancer therapeutics—from biotechnology companies and patients to investors and molecules—were facing an unknown future that was nevertheless fraught with hopes and expectations. Which research strategy should biotechs follow when biochemical mechanisms were unknown? On what grounds could they secure funding for their research expenditures?

And would the novel product pass through clinical trials and obtain regulatory approval? At the same time, competition between biotechs was intense as each sought to find a successful treatment while also striving to hit a financial jackpot. This search can be seen as a window onto not only the development of cancer therapeutics and the biotech industry at a pivotal time, but also more generally onto how innovations emerge and are transformed amid a nascent market of competitors.

The book traces the evolution of "innovative breast cancer therapeutics," beginning from the late 1980s until 2010, to better understand how market emergence and transformation occur in dynamic conditions of innovation. At the center of drug discovery, it shows, lie complex innovation processes in which scientists, biotechnology companies, and their funders and investors need to make sense of ambiguous findings and grapple with numerous and unpredictable interdependencies over many years of product development. To do so, they use stories, especially public stories of the future. Applying a combination of qualitative and large-scale text analyses to thousands of press statements, media reports, scientific reports, and financial and industry analyses related to innovative breast cancer therapeutics, *Making Sense: Markets from Stories in New Breast Cancer Therapeutics* identifies stories as the central mechanism for the emergence of markets and develops a novel methodology for theorizing emergent processes in the economy.

How actors make decisions in situations of uncertainty is central for the research field of economic sociology. In numerous studies it has been convincingly shown that social relations, institutional frameworks, as well as practices contribute to reduce uncertainty in decision-making processes. Actors' social and institutional embeddedness structure their expectations and actions in market contexts, thus enabling coordination. Typically, such analyses have dealt with existing markets, or they have looked at how markets emerged from an *ex post* perspective.

Here the inquiry is into a market—the innovative breast cancer therapeutics market—during its emergence and from a perspective *as it is emerging*. My argument is that in order to understand how actions are coordinated in situations when it is unclear who participates and who will profit and thus to understand market emergence, accounts of structural mechanisms are not sufficient. Instead, I argue for *a relational and processual perspective on markets as networks of sociocognitive and sociomaterial actors*. Such a theoretic model allows a focus on the narrative and evaluative dimensions in emergent processes.

The result is a cultural analysis of market emergence that shows its crucial elements and mechanisms in three phases over time.

This book posits that actors leverage culture to grapple with ambiguities of newness in the emergence of markets. A diversity of actors, such as companies, industry analysts, journalists, and scientists, tell stories in which they evaluate and interpret the present, the past, and a projective future. These stories then serve as cues for other market actors. Therefore, this book also suggests that meaning-making is crucial for market emergence.

An inquiry into the stories of potential members of a new market allows for the tracing of individual and collective interpretations and meaning-making. In turn, collaboratively these actors' stories participate in category construction on what the new market is about. Such an inquiry sheds light onto both successes and failures in category construction. The collaborative process of category construction will not be friction-free; indeed, rivalries and dissonances are conducive to entrepreneurship (Stark 2009) and, as a by-product, to category construction. Moreover, no individual firm is able to claim to have a worthy product or to be a worthy market participant without receiving recognition by others. Similarly, no product can claim to be a worthy market participant unless tested and proven so over time and unless other market participants recognize its worth. In the ambiguous setting of a nascent market, worth gets negotiated collaboratively in struggles and agreements.[2]

Through its empirical analyses of economic life from a cultural perspective, the book delineates how the interrelations between multiple network domains of science, biotech companies, and financial and industry analysts collaboratively come to form a market. The analyses build on the theoretical insight that networks are composed of "culturally constituted processes of communicative interactions" (Mische 2003) across heterogeneous actors. Empirically this translates into tracing *stories*. The stories selected as data for analyses originate from public statements of companies that claim to be involved in breast cancer therapy research, as well as from financial analysts, industry analysts, journalists, and the scientific community as published in specialized media outlets. This selection of data thus purposefully allows for a look at the entangled developments in breast cancer therapy research from different though interrelated perspectives at the time they were happening. Via a combination of qualitative text analysis and novel computational methods of large-scale text analyses, the book captures the dynamics of meaning-making processes over twenty-two years in what is now understood as a period of medical transformation toward targeted or so-called personalized medicine.³

One set of stories examines the state of the art of scientific findings. They provide answers to the questions of what topics research on breast cancer therapeutics has focused on and how that research has evolved. These stories offer many facts, but they also address hopes and expectations. Other sets of stories furnish insights into how actors evaluate their situation and context, and how they act accordingly. Narrated stories establish interpretations that serve as stabilizing forces for actors' identities. At the same time, actors are in narrative competition with one another about which story counts, how a situation should be interpreted, and what social formation they are part of. Dominant interpretations may come to be confirmed or contested. Moreover, studied over time, such sets of stories also show how interpretations can settle into categories, understood as narrative constructions with real material consequences. To be sure, the perspective taken here does not trace individual stories of individual actors, but rather understands stories and their observations as relational ties of a larger network of cognitive interdependence. As actors tell stories about an uncertain present and a projective future, they share "socially structured imaginaries" about what counts as worthy (Fourcade 2011) and provide *interpretations of themselves* that are also *for others*.

Two central findings result from the analyses of these sets of stories and their workings. First, the study finds that economic actors tell *stories of the future* and thereby create a *market of expectations*. Prior to the existence of any innovative breast cancer product, biotechs are trading stories on expectations, projections, and imaginings about what the future might hold. These stories provide cognitive guidance for the involved actors amid an ambiguous setting. Moreover, journalists and financial analysts similarly voice their bets about an uncertain future; they too face ambiguities of newness and need to cope with the many unknown unknowns related to how to interpret and evaluate biotechs' research reports. Hooking on to stories of the future, filled with expectations, promises, and hopes, helps all involved to cope with uncertainties of the situation. Together and with their entangled stories, these actors are creating a market of expectations.

Second, the study finds that the stories temporarily settle into categories that in turn have real-world consequences: they induce changes in the market structure. Biotechs' research strategies, their funding, and their positioning in the emerging market are influenced by the stories actors tell—and by the observation of their competitors. The study finds that competing actors are involved in collaborative meaning-making processes on how the new market can be categorized. The *construction of a new category* results from negotiations of newness and from dealing with ambiguities. It helps to, albeit temporarily, stabilize the market. New categories develop as constructs in which to store temporary associations as well as temporary imaginations of what constitutes actors worthy to connect to that category. In turn, categories provide a mechanism for inducing market structure.

These insights go beyond a simple "stories matter" message and instead provide support for the argument that narrative constructions play a fundamental role in the economy (Beckert 2016, 2021). The analyses show that decision-making in situations of great uncertainty, when it is unclear what the outcome will be, and who is competing with whom about what, is guided by stories of the future that the involved actors tell to direct and to justify their actions. Stories in this sense are the cognitive connections between a diversity of actors, which include organizations, people, tools, tests, and also molecules. Stories are the primary devices for contextual, situational meaning-making in markets and are thus microfoundations of decision-making. They are not "only talk." If they can inspire a belief in a specific future, they can motivate, orientate, and coordinate action. Moreover, stories are also instrumental in establishing categories of what the "new" is about. The analyses thus point to two crucial elements—expectations and categories—that rely on the existence of stories and that are necessary for the emergence of markets.

The study finds three phases in the emergence of this market. In the first phase, it is unclear what the products will be and who will participate. The phase begins in the late 1980s, when the "first biotech era" ends with the centerpiece biotech drug called interferon failing in cancer trials (Kaplan/Murray 2010; Pieters 1998, 2005). At this time, scientific advances suggested that a molecular, noninvasive treatment of breast cancer might be possible. Nevertheless, uncertainty about which research strategy to follow and which molecule to invest resources in characterize the field. The first phase ends when regulatory agencies approve the first product for breast cancer therapy based on molecular engineering in 1998.

A second phase begins after that first product approval. With the approval, market participants have a first product that can be used for comparison. Yet ambiguity about the products' mechanism and alternative research strategies persists. This is a phase of discursive struggles and contestations. Processes of categorization first move from a description that the market is about "innovatives" to identifying the market to be about "targeted therapies." This category, increasingly mentioned in self-descriptions and analyses, is able to stick with the heterogeneous actors constituting the market in the early 2000s. At that time, it has become scientifically evident that no blockbuster, no cure for all would be possible to develop for breast cancer. Indeed, the future of cancer research would be targeted treatments moving toward "personalized medicine" (Langreth/Waldholz 1999). The empirical analyses show how the new category *targeted therapies* stuck.

Once the category was taken for granted and institutionalized after much contestation, and only after the approved product had been collectively labeled to exemplify a new category of therapies, did the market transpose and diversify. This is when the third phase begins, in which more products get approved. Clinically, breast cancer is categorized into three basic therapeutic groups depending on hormonal and other receptors. Beginning with the early 2010, however, research on the molecular structure of breast cancer pointed to a more complex picture: breast cancer can be classified into four genetically distinct types, which with their combinatorial subtypes yield at least forty genetic variations to which targeted therapies may be developed (Cancer Genome Atlas Network 2012; Schnitt/Lakhani 2014; Sinn/Kreipe 2013). Mounting evidence on genetic mutations and results from clinical trials currently suggest that it is not suitable anymore to simply categorize cancers by tissue. Thus, after 2010 new categorization attempts can be found. Overall, the analysis traces the different attempts to make sense of products and scientific developments.

While the empirical focus is on a biomedical field at the overlap of science, commerce, and finance, the book's contributions fundamentally speak to ongoing discussions in sociology.

First, the book contributes to recent developments in economic sociology toward cultural processes (Beckert 2013, 2016, 2019; Wherry 2012, 2014) when it focuses on the role of stories to explain how markets emerge. It complements a solely structural perspective on the emergence of new social formations (Padgett/Powell 2012a) by taking sociocognitive and sociomaterial actors into account (Stark 2009). Focusing on economic actors, their setting, and their products as well as the stories that create and shape them, the book advances a perspective for a cultural analysis of market emergence.

In doing so, the book contributes to ongoing discussions on meaning-making processes. This contribution has two dimensions. One dimension is a substantive one in that it delineates categorization processes. US and European cultural sociologists and organizational scholars alike have become acutely aware of valuation and evaluation processes as basic social actions at stake when sorting, classifying, and categorizing occurs (e.g., Lamont 2012). In particular, the book focuses on the *How* of category construction in ambiguous situations.

Another dimension is a methodological one. Building on insights from innovation studies on how to study newness, the book takes a not only relational but also a processual perspective to study dynamics of market actors over time. To do so, the book engages with discussions on how to measure meaning (Mohr 1998) and employs a combination of methods. It combines in-depth qualitative text analyses with novel, computational methods of text analyses, i.e., topic modeling and semantic network analysis, to enable a sociological inquiry into meaning-making over time. Moreover, such a design permits us to overcome the often-faced divide between qualitative and quantitative approaches in cultural analysis and in this quest joins current research (e.g., Bail 2014; Breiger et al. 2018; Karell/Freedman 2019; McFarland et al. 2013; Mohr et al. 2020; Mohr/Bogdanov 2013; Mohr/Rawlings 2015; Mohr et al. 2015; Nelson 2020, 2021).

The content of the book bridges and recombines elements from different fields of academic discussions, including economic sociology, sociological theory, innovation studies, organizational studies, and methods of cultural analysis, to study the complex field of commercial cancer research as an example for the framework of markets that come from stories. Additionally, by describing the scientific developments in cancer research, the book intends to provide a non-expert readership with a basic understanding of the field. Similarly, computational methods used are introduced in nontechnical terms. Their technical details can be found in the Appendix.

In this introductory chapter, I first sketch how the problem of innovating in the field of biotechnology has been analyzed and point to temporal, relational, and cognitive complexity in innovation processes. I then introduce the larger theoretical perspective of the book and how that translates into methodological choices. In the last part, I lay out the organization of the book.

Innovating the Sociology of Innovation

The field of biotechnology and its complex story of fears, hopes, promises, risks, ethics, science, and business have been extensively studied by sociologists as well as scholars of organizations and of science and technology alike (e.g., Adam et al. 2000; Bauer/Gaskell 2002; Evans 2002; Kenney 1986a; McKelvey 1996; Pisano 2006; Rifkin 1998; Robbins-Roth 2000). Social historians, historians of science and technology, and scientists have described how cancer treatments have developed in the past (e.g., Chabner/Roberts 2005; DeVita/Chu 2008; Keating/Cambrosio 2012; Lacroix 2011; Mitchinson 2005; Mukherjee 2010; Timmermann 2014; Van Poznak et al. 2002); business historians and journalists have indicated the role of entrepreneurs in developing cancer treatment innovations (e.g., Bazell 1998; Hughes 2011; Vasella/Slater 2003). This book builds on these accounts and offers an alternative lens when it studies the evolution of molecular breast cancer therapeutics to identify important elements in market emergence given a field of innovations.

The study of innovations in the organizational and sociological literature on biotechnology has previously been shaped by three dominant perspectives. First, studies have highlighted the role of national innovation systems (Nelson 1993). They consider institutional governance structures to explain national innovative output (e.g., Bartholomew 1997). Such innovation systems determine the technological regimes that govern innovative activities (Breschi et al. 2000) and, in turn, the evolution of the biotech industry overall (Allansdottir et al. 2002; Garavaglia et al. 2012; Malerba/Orsenigo 2002). This perspective focuses on output in terms of innovative products, often using patent data or research and development (R&D) expenditure, to explain the structure and capabilities of the industry.

Another perspective focuses on the connections between a diversity of actors involved in the field of biotech. Such ties may serve as resources that help to create knowledge (Arora/Gambardella 1994; Gittelman/Kogut 2003); to adapt to organizational challenges (Maurer/Ebers 2006); to innovate at the level of individual firms (Swan et al. 2007), of project ecologies (Newell et al. 2008) or of entire industries (Galambos/Sturchio 1998; Pisano 1991); and to learn from others (Owen-Smith/Powell 2004; Powell et al. 1996). In sociology in particular, biotechnology has been extensively studied as an organizational field (DiMaggio/Powell 1983) with network analytic tools, looking at linkages between inter-organizational agreements (Owen-Smith et al. 2002; Powell 1996; Powell et al. 2005), co-inventors' patents (Almeida/Kogut 1999; Fleming et al. 2007), and inter-firm personnel mobility (Casper 2007; Casper/Murray 2005) to point to the role of communities and regional clusters (Whittington et al. 2009) and to shed light on firms' performance (Powell et al. 1999). This perspective indicates what types of organizational settings and which modes of connections between actors are necessary to make innovative outcomes possible.

Furthermore, others have emphasized the role of Schumpeterian entrepreneurs in their quest to profit from innovations (Kenney 1986b; Shane/Venkataraman 2000). This perspective views innovations as something to exist already, and which just need to be picked up by astute entrepreneurs.

In recent years, however, research on the evolution of technology has shown that neither a technology nor its organizational field is fixed and that entrepreneurs thus are not able to "simply pick up" good ideas (Garud et al. 2007; Garud et al. 2002). Rather, technology is socially and organizationally constructed (Bijker et al. 1989); it is mutable and shaped as the field and its actors evolve (Garud/Karnoe 2001). But if technologies, their fields, and their actors are in co-evolution, the study of innovations as outcomes misses its moving target. Instead, as this line of research suggests, we need to study unfolding processes to account for where newness comes from. The question then becomes: what are the mechanisms that can account for new things to arise?

Emergence is a classic theme of sociological theories. It is typically framed as a link between different analytic levels, e.g., in order to explain the existence of collectivities on the basis of individual behavior (e.g., Alexander et al. 1987; Knorr Cetina/Cicourel 1981). Some, like Durkheim, argue that collective phenomena are collaboratively created by individuals but are not reducible to individual action. Others, like Homans and Coleman, argue that collective phenomena can be explained by individual properties and their relations. Certainly, emergence has proven to be a "slippery concept" (Sawyer 2001), which has intensified disciplinary divisions due to prioritizing one analytic level, micro or macro, over another (Heintz 2004). Outside of sociology, and less concerned with the correctness of analytic levels, complex adaptive systems theory science has started to formally model emergent processes (Holland 1998; Johnson 2001). Emergence here refers to "the arising of novel and coherent structures, patterns, and properties during the process of self-organization in complex systems" (Goldstein 1999: 49). Emergent phenomena "have features that are not previously observed in the complex system under observation. This novelty is the source of the claim that features of emergents are neither predictable nor deducible from lower or micro-level components" (50).

Recently, sociological network studies have integrated these insights from complexity theory into their theoretical framework and empirical studies. In their work on the emergence of new organizational forms, new organizational actors, and new markets, Padgett and Powell have shown that newness depends on structural mechanisms of "network folding" (2012b: 12), in which practices or relations are recombined across different social domains. In turn, this may lead to spillovers across multiple, intertwined social networks, in effect completely changing how things are done. For example, in the 1970s, academic science labs recombined with the practices and relations of venture capital to become dedicated biotechnology firms, a novel type of organization at the overlap of science, commerce, and finance (Powell/Sandholtz 2012b). These new organizational, amphibious forms then led to spillover in the pharmaceutical field as they changed the nature of how biomedical research on new drugs was conducted. Some came to be "in business to do science," others "in science to do business" (Powell/Sandholtz 2012a). Padgett and Powell's endogenous model of newness takes processes across multiple network domains, rather than just individuals and their interactions, into account and specifies *structural mechanisms* for newness to emerge.

Recent works in organizational studies have picked up on the complexities of innovation processes and the emergence of newness from a different perspective (Dougherty/Dunne 2011; Garud et al. 2015; Van de Ven et al. 1999). These works look at innovation processes by observing the sequences of events that unfold over time, "events that unfold as ideas emerge, are developed, and are implemented within firms, across multi-party networks and within communities" (Garud et al. 2013: 776). This stream of work points to *temporal, relational, and cognitive complexity in innovation processes*. Innovation processes are most often not linear and they may entail failures, so an empirical perspective over time, as the process evolves, is necessary. History not only matters, but actors may experience time in different ways. Some may only have future orientations (Brown et al. 2000), while others try to make sense of the past (Weick 1995). Diverse sets of social actors and material elements are involved in innovation processes (Latour 2005). This stream of work also suggests that interpretative processes of multiple actors are involved in innovation processes to cope with uncertainties (Garud/Rappa 1994; Kaplan/Tripsas 2008; Schubert et al. 2013). For example, in the history of biotechnology, entrepreneurs contested interpretations of "what biotechnology meant at any one point" (Kaplan/Murray 2010: 108) and constructed different economic logics of the field in which they operated.

Such a "cognitive turn" in the study of innovation processes (and generally, in institutional theory, e.g., DiMaggio/Powell 1991; Schneiberg/Clemens 2006) parallels a "cognitive turn" in sociology (e.g., DiMaggio 1997; Zerubavel 1997). For one, and as part of the notion of "culture as a toolkit" (Swidler 1986), cognitive frames and cognitive boundaries came to be building blocks of a US sociology of culture that rejects "the assumption that culture generates values that drive social action" (Kaufman 2004: 340). Rather, it argues that "culture organizes thinking by providing the frames and schemas that individuals and their groups use to process and translate their environment" (Fine/Fields 2008: 136). This helped guide cultural sociology to being centrally concerned with meaning-making (Lamont 2000).

At the same time, some structural sociologists using social network analysis similarly began to focus on culture, not as a separate entity anymore, but as entangled with and co-constitutive of social structure (White 1992). What has become known as relational sociology is central to the turn of network analytic works toward the cultural, including practices, discourses, and stories, with a similar focus on meaning-making. Indeed, by now an increasing amount of research is spanning the boundaries between culture and networks (for overviews see DiMaggio 2011; Mische 2011; Pachucki/Breiger 2010). Key cultural sociologists argue for network analysis "as the natural methodological framework for empirically developing insights from leading theoretical approaches to cultural analysis" (DiMaggio 2011: 286). In turn, key network analysts consider how "culture prods, evokes, and constitutes social networks" as an integral part of their analysis (Pachucki/Breiger 2010: 219). These lines of work have shown that cultural meanings shape social structure (e.g., Lizardo 2006), and that social networks are also culturally constituted and interwoven with meaning (e.g., Mische 2003, 2008).

Framing the evolution of biotech's high-stakes quest for breast cancer treatments as a window onto broader processes of innovation and market emergence, this book builds on such a focus on meaning-making when it brings to the study of innovation insights both from relational sociology[4] and actor-network theory (ANT). Individually, relational sociology and actor-network theory each provide distinct approaches to study temporally, relationally, and cognitively complex processes of how sets of actors co-evolve to produce newness and how meaning-making comes about, by connecting, observing, and narrations. While there are certainly similarities and differences between these two analytical approaches (see e.g., Mützel 2009), this book argues for an expansion of structuralist mechanisms by taking sociocognitive and sociomaterial processes across a multitude of actors into account to explain processes of innovation and the emergence of a new market. At its core lies a theoretical perspective that expands a predominantly structuralist relational sociology to also take objects and concepts together with humans and organizations as actors into account. It thus imbues relational sociology with key aspects from actor-network theory.

Fundamentally, relational sociology sees "relations between terms or units as preeminently dynamic in nature, as unfolding, ongoing processes rather than as static ties among inert substances" (Emirbayer 1997: 289).[5] Moreover, it rejects "the notion that one can posit discrete, pregiven units of analysis such as the individual or society as ultimate starting points of sociological analysis" (287).[6] Central reference for such a relational sociology in which network and cultural thinking intersect are the programmatic writings of Harrison White (1992, second ed. 2008). They present a reconceptualization of how we understand actors, action, and social relations, by analyzing how identities, relations, and their social formations, with all their "target and content ambiguities" (Leifer/Rajah 2000: 252), emerge.

White starts off with the observation that we live in a world of contingencies and social chaos, which we as social actors are able to maneuver, because we are able to couple and uncouple social ties across multiple social contexts. How so? White analytically conceives of *identities*, i.e., pre-person actors, to emerge amid contingencies and contention in interaction with other identities in attempts to establish *control*, i.e., their social footing in relation to one another. To establish control and hence social footing, identities use discursive interactions, to which other involved identities attribute meaning. These discursive interactions, or *stories*,[7] are open for interpretation and are directed at a plurality of listeners. In telling stories, identities thus establish polymorphous relations.

These efforts to establish social footing occur in *network domains* (*netdoms*) that are realms of interaction characterized by bundles of relations and associated sets of stories, entangled with each other. In netdoms, structural and cultural dimensions of relations coalesce. Moreover, netdoms have a temporal dimension: since each type of tie is accompanied by stories, which relate to the past or future, the relational tie provides for temporal extension. It is in such meshing of culture and structure that identities relate to and are in struggles with other identities. From the point of view of identities, they continually switch across netdoms in search of social footing in their everyday life. Such situative *switching* entails a temporary uncoupling from an old and coupling with a new netdom. Switching between netdoms allows identities to engage in reflective comparison and, consequently, collaboratively generates new perceptions and meanings (Mische/White 1998; White 1995, 2003; White et al. 2007; White/Godart 2007).

Building his approach on empirically observable uncertainties and contingencies in action encountered in everyday life, White departs radically from the static orderliness of structuralist network thinking. In order to capture these fleeting processes of actor-actions interrelations and cross-constitution, he suggests to study action in its actual context, from the point of view of processes, and "how they unfold together over time" (1992: 78). Thus relational sociology understands networks as composed of "culturally constituted processes of communicative interactions" (Mische 2003: 258), providing for an inseparable intermingling of network relations and discursive processes. Culture and structure, language and relational ties are *fused* within a sociocultural setting.

To explain processes of innovation and emergent social formations, this book argues that relational sociology needs to expand its structuralist core by taking sociocognitive and sociomaterial processes across a multitude of actors into account. This translates into a wider notion of who and what counts as an actor and a methodological approach that follows the actors as they connect with, relate to, and observe others. Key insights from actor-network theory present additional building blocks for the book's theoretical foundation.

For actor-network theory (ANT) networks are a heterogeneous chain of associations made up of multidimensional and evolving entanglements of human, nonhuman, or collective actors—all are *actants* (Callon 1986a, 1998b; Latour 1986, 1999, 2005; Law/Hassard 1999). The analyst's task is to follow the ways in which actors link up with other actors through activities. Such linking up occurs in the process of *translation* that analytically separates into different moments (Callon 1986b) and that is key for the study of meaning-making and the emergence of newness. Translation involves that a complex situation, in which many entities are involved, needs to be transformed into a well-defined problem: potential allies need to be identified, they need to become interested in involvement, and eventually they need to be enrolled and to mobilize support for definitions and further entanglements. In the process of translation, actors and their meaning get defined and redefined collaboratively. For example, a microbe may not only be a health hazard; it may also help to establish new hygienic measures and enhance individual actors' professional careers (Latour 1988). Translation makes newness possible.[8]

The key methodological perspective is that analysts follow the actors and their constitution of categories instead of defining categories *a priori*. In this way ANT-based studies present accounts, in which actants are traced in how they accomplish the processes of establishing associations, i.e., constructing persons, groups, and the social—and thus offer descriptions as explanations (Latour 2005). Analytical focus is first on the multifaceted interconnections of a local, egocentric network of an actor, before moving to the next connected local bundle of entanglements. Eventually these shifts and redefinitions between one micro-network of associations to the next over space and time add up to a larger narrative on transformations of ideas and practices. These aspects allow accounting for the emergence of new social actors (e.g., Latour 1988) or categories (e.g., Bowker/Star 1999), especially in the field of technological and scientific innovation.

Another methodological element of ANT studies that proves fundamental for a fruitful expansion of relational sociology is the use of large textual corpora that visually establish reified relations between heterogeneous actors, such as concepts, institutions, molecules, scientists, and diseases (e.g., Bourret et al. 2006; Callon 2006; Cambrosio et al. 2006; Cambrosio et al. 2004) or terms and concepts in scientific controversies (Venturini et al. 2014). These works thus extend early ANT work on co-word analysis (e.g., Callon et al. 1983) and, similarly, network analytic studies on concepts (e.g., Carley 1997; Mohr/Lee 2000; Schnegg/Bernard 1996), categories (e.g., Martin 2000; Mohr/Duquenne 1997), and narrative clauses (e.g., Bearman et al. 1999; Bearman/Stovel 2000) treated as nodes. Such studies offer *composed configurations*: "As the relations visualized are real, the analyst is able to see all the dependencies woven by the different actions" (Callon 2006: 11); affiliations between authors, keywords, and institutions can show profiles of concepts (Latour et al. 2012).[9]

This book's case study of breast cancer therapeutics extends relational sociology in two important ways to provide for analytic leverage when studying processes of emerging social formations. It considers humans, objects, and concepts as equally able to act. In this process of emerging social formations at the intersection of science, business, and finance, all sorts of actors come to be relevant and create meaning: in clinical tests, molecules may not yield what they suggested in prior tests. In the case of an emerging social formation, companies who claim themselves to be developing innovative breast cancer therapeutics are considered to participate in social formation. Some may end up developing their biochemical mechanism further in a breast cancer product. Others may discover that it does not work against tumor growth after all.

Moreover, the study of individual innovations is not of primary interest. Rather, the focus of the book lies in tracing the processes of innovating across multiple actors and across multiple network domains. Tracing such processes allows finding contests over meaning to occur not only with an egocentric focus on one actor and its contextual relations, but rather across the entire network. This yields methodological consequences for the empirical operationalization. For one, the book makes use of qualitative data. Such data typically offer accounts on where the actors are coming from, where they are going to, and whom they take as reference in their social settings. The book examines thousands of documents, in which actors speak about themselves and others at the time when they act (their *stories*). This is a methodological attempt to cope with the fact that ethnography of an emerging field of the past is unfeasible and that interviews about *ex post* events would introduce hindsight, unavailable at the time of the actual doing. My interest lies in the tracing of meaning-making from available documents, as they were occurring across a multitude of actors in a particular context. The analyses use textual data produced by the biotech companies themselves, financial and industry analysts, journalists, as well as scientific reports. Moreover, the analyses extend both relational sociology's and ANT's methodological toolbox: they combine qualitative, interpretive methods with the power of formal relational methods to concomitantly capture cultural and structural aspects using topic modeling (e.g., DiMaggio et al. 2013) and semantic network analysis (e.g., Chavalarias/Cointet 2013; Venturini et al. 2014) on large-scale data sets over time to measure meaning-making (e.g., Mohr 1998).

Organization of Book

In hindsight, the emerging market of innovative breast cancer therapeutics can be delineated into three phases. In the first phase, it is unclear what the products will be and who will participate. In the second phase, we see approval of a first product for breast cancer therapy based on molecular engineering and the categorization of *targeted therapies*. In the third phase, the category *targeted therapies* is taken for granted and the emerged market for breast cancer therapeutics diversifies. However, at the time actors were making decisions to follow research strategies and to develop further molecules to turn them into possibly lifesaving biomedical innovations and eventually into market products, this narrative was of course not available. It is only through analyses that these phases and pivotal turning points of developments come into focus. The organization of the book follows the timeline of developments and takes close and distant analytical views onto the stories of market emergence.

Chapter 1 introduces the theoretical model and fleshes out the perspective of markets as networks of sociocognitive and sociomaterial actors, in which actors tells stories. First, the chapter discusses different perspectives on how markets are constituted and how actions are coordinated. Throughout the discussion of perspectives, the chapter indicates how the perspectives relate to each other. While these perspectives highlight the structural, cultural, relational, cognitive, material, and evaluative elements pertinent in analyzing markets, their focus is typically on already existing markets. The chapter then inquires into works that address market emergence and highlights dimensions, processes, and mechanisms thereof. It links a structural approach on the emergence of newness to one that takes reflexive cognition, interpretation, and evaluation practices into account. These elements of cultural processes taking place in markets lead toward the second section of the chapter. This section turns to the role of stories as the central mechanism for the emergence of markets. It examines several approaches on the study of communication of economic and organizational actors. A third section discusses how stories of a projective future shape market-making processes. This chapter purposefully serves as a framework for discussions both on markets and on stories to stress their interconnectedness, which ultimately comes together in the analytical model and explanatory theory captured in *markets from stories* that guides the rest of this book.

Chapter 2 introduces the empirical context: breast cancer and the search for therapeutic treatments. This chapter provides a basic overview of breast cancer and how it can be treated. After an introduction to the case of breast cancer, it presents a brief history of the treatment of breast cancer, including crucial developments in oncological and molecular research, up until the late 1980s. The second section describes the field of biotechnology up until the late 1980s.

Based on a qualitative text analysis, Chapter 3 traces how actors in the field of "innovative breast cancer therapeutics" are trying to make sense of what they are about and with whom they are competing by examining their stories.

The analysis highlights ambiguities about newly developed molecules as well as biotechs' research strategies. The chapter shows how the involved actors are cognitively interrelated and come to collaboratively construct a market of expectations, which, at least temporarily, turns out to be a profitable strategy.

Chapter 4 shows how different socioeconomic actors come to make sense of the market once a first therapeutic product is approved. The inquiry focuses on market research analysts' stories, their framing of the future, and how to evaluate new products. It delineates categorization processes. After some disputes, the category *targeted therapies* is established as that which the market of this new type of therapeutics is about. This category is collaboratively constructed across multiple network domains of science, finance, and business.

The empirical analyses of Chapter 5 then take a macroscopic view of the global market over twenty-one years. It is based on large-scale textual corpora from scientific discussions, newspapers, and press release data. Using computational methods to analyze large textual corpora, this chapter shows how the category *targeted therapies* comes to be institutionalized in all four field of inquiry, thus promoting the institutionalization of a new market of targeted cancer therapeutics. Topic modeling analyses highlight different field-specific trajectories. Semantic network analyses show the shifts and drifts of research strategies and of targeted therapies in oncological research over time. From a global perspective this chapter shows the construction of the new scientific and then also economically viable category *targeted therapies* as formative for the new market for breast cancer therapeutics.

The Conclusion highlights findings of the empirical inquiry and relates them to my analytic model and explanatory theory. It indicates its contributions to ongoing discussion in the field of sociology and suggests how the approach that stories are the central mechanism for the emergence of markets applies more generally to the study of emerging social formations and points to areas of future research. I conclude with a look at developments in the field of cancer research, which continue to stir high hopes for millions of patients.

1 Markets and Stories

IN THEORIZING AND THROUGH empirical study of how market emergence and transformation occur in dynamic conditions of innovation, this book addresses a foundational question for economic sociology: where do markets come from?

Ever since White asked this question (1981), economic sociologists have been studying the social, non-economic factors involved in the constitution of markets in situations of uncertainty. Separately or in some combination thereof, empirical research over the past four decades has identified the role of social relations, institutions, politics, morals, culture, cognition, calculative devices, objects, and concepts in the constitution of markets. Typically, however, such analyses deal with already existing markets or they look at how markets came into existence from an *ex post* perspective. By inquiring into a market that is just beginning to emerge from a perspective *as it is emerging*, this study offers a new methodological approach to studying where markets come from that leads to a new set of arguments about how actors make decisions in situations of uncertainty.

The central argument this study puts forth, summed up in the phrase "markets from stories," is that stories—and especially public stories about the future—are the central mechanism for the emergence of markets *and* the portal through which market emergence and transformation can most effectively be studied. My argument is that in order to understand how actions are coordinated in situations when it is unclear who participates and who will profit and thus to understand the emergence of markets, accounts of embedded ties as well as structural mechanisms are not sufficient. Rather, to answer the question "Where do markets come from," I advance a perspective on *markets as networks of sociocognitive and sociomaterial actors*, and study them in a relational perspective over time, to capture the evaluative and narrative dimensions in emergent processes. I thus add to contributions arguing for combined approaches (e.g., Beckert 2010; Fourcade 2007) to better understand dynamics of markets and their microfoundational processes.

An analytic model and explanatory theory, markets from stories builds on previous research from multiple fields, which it brings together, as well as on the empirical study of innovative breast cancer therapeutics. The focus of this chapter is on the former. Previous works in economic sociology, as the first section discusses, have theorized market as social formations by identifying the interrelated cultural, relational, cognitive, material, and evaluative elements that contribute to markets and are pertinent to analyzing existing markets. Yet these approaches, I show, have not adequately accounted for the dynamics of emergence, which are necessary to understand nascent markets and the processes through which such emergent social formations become stabilized. As the chapter's next section explains, it is necessary to integrate the analysis of stories in theorizing emerging processes, because it is through stories that actors navigate uncertainty, settle ambiguity, and coordinate action to produce markets. I argue that we need to view markets as networks of sociocognitive and sociomaterial actors and focus on the stories of the market. Furthermore, I show that to focus on stories of the market, it is necessary to extend previous inquiries to encompass the role of public stories of third actors, fragmented narratives, stories of the future and their projectivity, as well as on the role of ambiguity and how categories help to temporarily settle ambiguity. Having brought together current perspectives of markets as social formations and the importance of stories to theorizing emergent processes, the final section of this chapter more concretely articulates the analytic model and explanatory theory of market from stories, which will receive further development and application in the subsequent empirical study of innovative breast cancer therapeutics.

Markets as Social Formations and the Dynamics of Emergence

A central, self-defining tenet of economic sociology is that markets are social formations based on structural relations. One fundamental perspective holds that a relational tie between two economic actors functions as a channel or "pipe" (Podolny 2001). According to this perspective, which follows the tradition of structuralist sociological network analysis, various resources such as information, money, friendship, or advice flow through these channels and thus connect actors with each other. Seminal works following this perspective understand economic relations shaped by social relations, often with repeated exchanges. Market actors are embedded in interpersonal networks that are trustworthy, provide information for economic opportunities, and shape market outcomes (Abolafia 1996; Baker 1984; Burt 1992a, 1992b; Granovetter 1985; Mizruchi/Galaskiewicz 1993; Uzzi 1996, 1997). This perspective of *networks as embedded in markets* has been widely influential for the study of organizations and markets in sociology, organizational studies, and economics alike (e.g., Granovetter 1995; Gulati/Gargiulo 1999; Jackson 2008; Rauch/Hamilton 2001).

A second perspective views *networks as alternative forms of governance* to firms and neoclassical markets (Powell 1990). As a "network form of organization" such market relations include "a wide array of joint ventures, strategic alliances, business groups, franchises, research consortia, relational contracts, and outsourcing agreements" (Podolny/Page 1998: 59). These network forms of organization are thus also based on direct exchanges.

These two perspectives, which focus on structural relations between economic actors and which often use network analytic tools, were instrumental in establishing the sociological study of economic action in US sociology in the 1980s. With the cultural turn in US sociology that moved the discipline away from Parsons's view of culture as "internalized value orientations" and toward the study of culture as practices and meanings, however, the study of *economic relations turned cultural*. New institutionalists in organizational analyses indicated the need to study "taken-for-granted scripts, rules, classifications" (DiMaggio/Powell 1991: 15) and "such standardized cultural forms as accounts, typifications, and cognitive models" (27) of relations in firms and across firms. Markets and their organizations have their own institutional logics on which economic decisions are based (Thornton et al. 2012; Thornton 2004). Culture thus becomes a source of strategic action (Fligstein 2001; Fligstein/McAdam 2012). Others showed how cultural meaning is implicated in economic relationships. Material things as well as the economic resource money may have different meanings attached, depending on who it was exchanged with and for what purposes (Zelizer 1978, 1989). To be sure, these meanings and interpretations attached to economic actions and material objects are an integral part of interactions and are indeed constitutive of social relations (e.g., Bandelj 2012;

Zelizer 1996). They do not belong to some higher order or symbolic system. Yet again others showed how actual market exchanges are organized by shared cultural understandings and rules (Abolafia 1996; Geertz 2001). Market transactions, this perspective has shown, are thus not neutral but infused with moral aspects.

Another structuralist stream of work considers cognitive aspects and their patterns as decisive for the constitution of markets. The idea of the perspective *networks as lenses* is that the observable relations of two actors are signals for third actors to draw conclusions about the status of the actors involved and, in addition, to better understand the structure of the market and their own status (Podolny 2001). Status in this sense is not the same as quality, but rather its perceived proxy. Based on the cognitive perception of relational patterns, this line of work has shown how actors use status as a signal in order to reduce uncertainty in decision-making processes (Podolny 1993, 2005). In addition, producers are trying to use ascribed status strategically: Benjamin and Podolny show (1999) that wine producers attempt to switch to higher-quality segments of the market by affiliating themselves with higher-quality appellations. Often they fail because they depend on the evaluation of others, such as wine guides. Thus status is constituted relationally and over time and changes when markets relations change (Aspers 2006, 2009, 2010, 2011). Recent work has shown that status can have advantageous but also detrimental effects on an actor's performance (Bothner et al. 2012).

The status order perspective builds on White's early market model (1981). Its central idea is that producers can acquire information about the structure of a market by observing a small group of competitors who consider themselves equivalent and are thus comparable to each other. This information about the structure of market, gained by observation, allows each competitor to find a role and an unoccupied niche in the market. "Producers' basic concerns are to hold on to distinctive positions in their markets" (White 1993: 162), and, in turn, to temporarily suspend direct competition with others.[1] "Each producer defines its role in terms of the similarities and differences it has with respect to other producers" (White/Eccles 1987: 984). By observing competitors' behavior, as well as the signals they provide about their performance in terms of volume and cost, the quality of their products, their profits, and the strategy of their firm—for instance, in reports detailing their economic results, advertising budgets, or even personnel decisions—market actors acquire clues as to how to act themselves, while they base their decisions on their observations of the existing market. An actor's behavior, following the observation and evaluation of the competitors, provides the competitors, in turn, with recursive clues (Leifer/White 1987). Thus the market structure itself helps the participants to orient themselves. The observed market structure also guides decisions coping with an unknown future. Subsequently the prior market structure will be reproduced (Leifer 1985: 443). In this model of *markets as networks* a market gets co-constructed with rivals.

White's producer market model zooms in on the most important observation partners in the process of finding a market niche: competitors who are deemed as comparable. White speaks of a *one-way mirror*, "through which the producers cannot see the miscellaneous buyers, even though these buyers can see the producers" (White 2002: 34). To help producers better understand how consumers perceive the quality of their products and those of the competition as well as their position vis-à-vis others, the market research industry has developed.[2]

According to White, markets are the structured results of a tangled, self-defined clique of producers whose network-like connections are created over time through the observation of signals and their comparisons:[3] "Markets are tangible cliques of producers observing each other" (White 1981: 543). In other words, markets are the consequence of network structures that adhere to a logic of co-constitution—they are neither an indiscriminate group of firms vying for customers nor are they dyadic patterns of exchange between economic players. It is only possible for the market to reproduce itself if producers can find their niche, maintain it, and develop it. This is how producers defend themselves against the competition and reduce their uncertainty. Because of cognitive limitations, "stable markets cannot contain more than a dozen or so producers" (White/Eccles 1987: 984).

White's market model offers an alternative to classical economic theory. Unlike in classic economic theory, economic actors are not equipped with the necessary information for economic maximization and efficiency. Indeed, asymmetric information and unclear quality properties produce uncertainty for economic actors (Akerlof 1970). To reduce this uncertainty, economic actors look for signals from their competitors that could help to interpret competitors' interests, future strategy, and the quality of products. On the basis of such signals, actors can form their own market profile on production volume, price, and quality of the competitors.

White relies here on Spence's signaling theory (Spence 1973, 1974, 2002), while criticizing and expanding it. Developed for the case of labor markets, signals are "activities or attributes of individuals in a market that by design or accident, alter the beliefs of, or convey information to, other individuals in the market" (Spence 1974:1). While Spence developed his model with what Granovetter (1985) calls "atomized," not otherwise connected workers and employers, White enhances the idea of signaling with the central, structural insight that job applicants are positions within a long chain of producers and consumers from different but interrelated markets (White 2002: 105). This multiple connectedness of workers, companies, and markets provides the structural foundation to understand signaling to occur in situational contexts, in which no more than a dozen actors interactively construct the market they are involved in (31). White's market perspective is thus situated in contexts of observations and reproduction of market structure; at the same time, a market structure is always about multiple networks intercalated.

In White's model of a producer market, producers sort themselves into different markets. The evolution of markets "is the effort of each producer to differentiate its offering from those of the other producers. In doing so, it seeks to appeal to a specific segment of the buyer side" (White/Eccles 1987: 985). Producers know who their competitors are and thereby define the product. This also defines which producers will be outside a particular market and belong to another one. "The offering of the producers in the market are considered more similar to each other than they are to offering of producers not in the market" (White/Eccles 1987: 985). White's market model specifies one social formation according to his general model of *Identity and Control* (1992). In markets, identities struggle to find footing amid uncertainty. In order to compare themselves and find a position, they observe others. While producers collaboratively construct their market, a market, in turn, is connected to other markets (and their producers), in which producers have different positions in the market profile.[4]

In yet another perspective, which fundamentally disagrees with the idea that network relations are embedded in markets, advocates argue that markets are embedded in economics (Callon 1998a). The idea here is that economics itself participates in the construction of markets, as markets are planned and built following economic principles (Garcia-Parpet 2010). "Economics does not describe an existing external 'economy,' but brings that economy into being:

economics performs the economy, creating the phenomenon it describes" (MacKenzie/Millo 2003: 108).

Moreover, this perspective also disagrees with approaches that only consider purely social aspects of market construction. Rather, *markets are sociotechnical arrangements* (Çalışkan/Callon 2010). Actors take things, such as material objects, tools, and formulae, into account by interpreting them, making judgments, and performing calculations. Developed out of actor-network theory, this perspective then views markets as being organized as traceable, networked contributions of human actors, such as traders, buyers, sellers, who are engaged in valuation practices, as well as nonhuman actors, such as technical, material, textual devices, rules, economic formulae, and infrastructures that are being valued and judged (Callon 1992, 2007a, 2007b; Callon/Muniesa 2005; MacKenzie/Millo 2003). Human actors in such markets use distributed cognition and calculative capacities; Callon refers to this notion of actors as "homo economicus 2.0," as an alternative to the classic homo economicus or homo sociologicus (2008). This approach to markets has been most prominently used in empirical studies of financial markets (e.g., Lepinay 2011; MacKenzie 2006; MacKenzie et al. 2007).[5]

According to the research program for the study of *marketization* processes, which are at the core of markets' functioning, markets are not fixed entities. Instead, they are *market agencements*, in which innovation is the driver of competition, indeed the essence of competition and inseparable from market activity (Callon 2015, 2021). To study marketization processes entails the study of the architecture of markets, i.e., the structure of property rights, the human and nonhuman constituents, as well as the organization of competition, i.e., how goods are designed, how they become valuable, how they get priced, and how goods and people get mutually attached. The qualification of goods results from collaborative activity, of which market transaction is only one type (Akrich et al. 2002). This research program goes beyond the initial insight on the performativity of economics, in that it broadens it to specifically include the role of knowledge and materiality.[6]

Another stream of work that relates to both White's model on markets as well as to actor-network theory is that of the *économie des conventions* (EC) (e.g., Biggart/Beamish 2003; Diaz-Bone 2011; Favereau/Lazega 2002). Since the 1980s, EC has evolved as an interdisciplinary field of research involving economists, historians, and sociologists to address how actors coordinate with other human actors, objects, and concepts in uncertain and complex situations. As opposed to standards or external rules, EC understands conventions as shared logics of interpretation that serve as resources in the coordination of actions and thus in dealing with uncertainty. Conventions are principles of action that are formed when actors interpret their own and others' actions in critical situations; they are not ad hoc agreements.[7] In critical situations, actors voice opinions, give justifications for their arguments and actions, and involve other actors and material objects. Such disputes are grounded in disagreement about the relative worth of actors and objects. When actors argue, they bring together different actors, facts, and objects to support and to justify their claims according to particular, incommensurable principles of equivalence and according to general principles of evaluation (Boltanski/Thévenot 1999, 2006).

Based on different empirical material, e.g., how-to guides for managers and firms (Boltanski/Thévenot 1999, 2006), readers' letters to the editors (Boltanski 1996), documents of an environmental movement (Thévenot et al. 2000), these studies show that principles of evaluations are collectively shared orders of worth.[8] The establishment of collective conventions of equivalence helps actors to coordinate because they reduce cognitive work in critical situations.[9] When one is trying to establish others' worth, objects and devices necessarily help human actors to calculate. Objects and devices are used as referents on which reality tests or trials can be based, which in turn are essential in providing proof, to end uncertainty and possibly the dispute itself.

In recent years, scholars have built on these perspectives and extended them significantly. One of the arguments has been that singular perspectives will not suffice to explain markets or economic actions. Instead, perspectives that interrelate different aspects are needed to grapple with dynamics of markets and a diversity of actors. Fourcade (2007) has scrutinized how different French and US approaches to studying markets really are and has suggested possible synergies. Beckert (2010) suggests a simultaneous inclusion of social networks, institutions, and cognitive frames into an analytical framework to study markets as fields. Stark (2009) fuses together French and US approaches in developing a framework to study processes of valuation.

This book picks up such a quest of synthesizing analytic approaches to better explain the dynamics of markets, considering a diversity of actors. The analytical framework of the book builds on White's model of markets as networks and recombines it: it not only takes sociocognitive actors but also sociomaterial actors into account. But to understand the *emergence* of a market, the perspectives discussed so far, even in their recombination, require the addition of considering the unique context of market emergence.

While the perspectives already discussed highlight the structural, cultural, relational, cognitive, material, and evaluative elements pertinent in analyzing markets, their focus is typically on already *existing* markets. The following part thus inquires into works that address market emergence. It also serves as a gateway: it links back to prior discussions while also expanding perspectives significantly toward the study of processes; it delineates processes, and mechanisms of market emergence; it introduces further elements to be used in building the analytical model of the book. At the same time, the discussion shows that one crucial element is missing in these views on emergence, namely stories, which will be picked up in later sections. Here the focus is on two frameworks that offer a perspective on the emergence of newness, including markets.

In their framework on emergence of new organizational forms and new markets, John Padgett and Walter Powell (2012a) build on White's model of action in suggesting a market to consist of multiple intercalated network domains. Their framework stresses the role of *process* and suggests a way to cope with different temporalities. It is attentive to the time frames in which analyses are conducted and in which actors are viewed. For them, "in the short run, actors create relations; in the long run, relations create actors" (Padgett/Powell 2012b: 2). This mantra indicates that when viewed in short time frames, all actors and objects appear fixed and neatly bounded. When viewed over longer processes of time, however, "transformational relations come first, and actors congeal out of iterations of such constitutive relations" (3). They thus argue that sociologists need to "search for some deeper transformational dynamic out of which" actors emerge (3) and thus study processes.

New organizational forms, new markets, new actors, new objects, i.e., newness in general, do not emerge *ex nihilo*. Rather, new forms result "from combinations and permutations of what was there before. Transformations are what make them novel" (2). Newness spins out of endogenous processes and such processes can be modeled and analyzed when studying multiple, intertwined network domains; say, for example, of political, economic, and personal relations and how they change over time. Networks between positions and, in White's language, identities, need to be viewed as processes and not as structures.

The idea of newness Padgett and Powell advance is twofold: New organizational actors or markets emerge from the recombinations of people, practices, and relations across multiple social network domains, by which networks fold together in new ways (Padgett/Powell 2012b: 6, 12). This results in innovations, which "improve on existing ways of doing things" (5), e.g., a new purpose for an old practice. These network foldings may then lead to spillover effects in so-far-unrelated network domains, which may possibly lead to new practices and relations altogether through catalytic processes. This is what Padgett and Powell refer to as inventions, which "change the ways things are done" (5). Conceptually, these ideas on the interrelatedness of different domains and their co-constitution leading to network folding and spillover effects pick up White's concept of multiple networks and combine it with Ron Breiger's notion of the duality of persons and groups (1974). The latter formalized Simmel's idea of the intersection of social circles. Networks from different social domains are interconnected by the members they share. Dually, in individuals, different networks they belong to intersect. Padgett and Powell add to those two fundamental ideas of social structure a processual perspective. An organization is the reproduction and recombination of persons and rules, which actors from different networks bring along. In turn, people are collections and results of prior networks and their interaction rules. "In other words, both organizations and people are shaped, through network co-evolution, by the history of each flowing through the other" (Padgett 2012: 171).

This model on the emergence of organizations and markets offers a complex structuralist answer to how newness emerges with a focus on *endogenous processes* and *transformations* based on relational ties of actors in the making. This framework suggests studying innovations no longer as mere outcomes but as processes. However, the suggested types of analyses take place *ex post* with a set outcome in view and do not address the role of cognition, interpretation, and evaluation processes necessary to bring about transformations.

Extending these structuralist insights, in turn, David Stark's work shows the role of reflexive cognition, interpretation, and evaluative practices in ambiguous and uncertain situations for the emergence of newness and thus adds sociocognitive and sociomaterial dimensions. Stark (2009) offers a framework that takes *cognition* and *interpretation* into account: it embraces US philosophical pragmatism in the tradition of Dewey, French pragmatist sociology of Boltanski and Thévenot, as well as actor-network theory to shed light onto how actors make decisions in uncertain situations and how newness emerges.

Stark differentiates between problem solving and interpretation. In the mode of problem solving, actors have identified a problem to tackle and can come up with solutions on how to approach it. In the mode of interpretation, actors are involved in "open-ended, unpredictable conversations" and seek "spaces of ambiguity since the challenge is to integrate knowledge across heterogeneous domains" (Stark 2009: 3). The mode of interpretation is "a kind of search during which you do not know what you are looking for but will recognize it when you find it" (1). The ability for "reflexive cognition" (4) is a first requisite. Analytically, it breaks down into different moments. Actors need to be able to recognize and identify something, e.g., a new product or practice, by making so-far-uncommon associations from across different network domains. Another step involves giving the new thing a category to belong to and to which others may attach their own interpretation. In yet a further step, others need to be convinced of the new thing, so they can support it and its categorization.

These moments of transformation based on reflexive cognition thus expand Callon's notion of translation (1986b) with Boltanski's and Thévenot's principles of evaluation (2006). To recall, Callon introduces translation as the decisive process that transforms a complex situation into a well-defined problem. Potential allies, including humans, objects, and concepts, have to be identified, they have to become interested in involvement, and they eventually need to be enrolled and to mobilize support for particular definitions and entanglements. Stark too emphasizes situational uncertainty and ambiguity. But rather than focusing only on the establishment of associations, Stark specifies further *how* these associations are possible to achieve. It is not merely about observations that establish connections, as in White's case, and is not merely about mobilization as developed by Callon and Latour. Rather, these processes entail *interpretation* and *evaluation*. When actors come to make new associations, they observe, compare, calculate, interpret, and evaluate—not necessarily and not always in that order.

Orders of worth or conventions in the more general sense are a way to reduce uncertainty. However, Stark argues, this approach "cannot eliminate the possibility of uncertainty about which order or convention is operative in a given situation" (15). Indeed, entrepreneurship exploits this kind of uncertainty: "Entrepreneurship is the ability to keep multiple evaluative principles in play and to exploit the resulting friction of their interplay" (15). This happens at the *overlap of evaluative frameworks*. Friction between multiple evaluative principles, and hence rivalry and disruptions about what counts as worthy, are necessary for the generation of newness. Ambiguous and indeterminate situations with diverse performance criteria are thus valuable. They are rife with possibilities for interpretation and entrepreneurial action. Instead of consensus and shared understandings, dissonance and generative friction are valuable for newness to emerge and for entrepreneurs to profit. In other situations, we may find moments of "cognitive interdependence," i.e., islands of resonance amid a sea of dissonance when actors observe each other for cues on how to evaluate an uncertain situation (Beunza/Stark 2012).

Yet the question remains: how do market participants take others' judgments and evaluations into account? How do analysts get at narrations of valuation? In order to establish equivalence, and thus an idea of whom to watch for cues and to compare against, whom to evaluate and interpret, in situations of uncertainty, market participants are involved in the telling of and the listening to stories. Stories narratively connect a diverse range of human and nonhuman actors. Stories are involved in making associations between firms and products and in establishing comparability or incommensurability. Stories also help to interpret and evaluate. Nevertheless, an evaluation of their meaning and also an assessment of which stories count when and for whom is not obvious and may be contested.

I maintain that both frameworks, i.e., Padgett and Powell's structuralist framework on the emergence of newness and Stark's important extension of reflexive cognition, complement each other. I will use their fusion as fundamental for the analytical model of this book. At the same time, this perspective's emphasis on interpretation and evaluation leaves open on what basis actors can interpret and evaluate. The perspective that I promote is that it is *stories* that are the primary devices for interpretation and evaluation. Stories need to be added to these two frameworks of emergence.

Stories and the Making of Categories

This section picks up the relevance of stories for describing and analyzing markets and thus adds another element to the prior perspective on market emergence. I argue for *stories* as primary *devices* for contextual, situational meaning-making in markets. Stories in this sense are not "only talk" but motivate, orientate, and coordinate action. To recall, White (1992) maintains that it is *stories* that are the ties, which connect actions with action, action with actors, and actors with actors. Stories accompany the emergence of identities from action and counteraction as they are "generated by control efforts" (13), "soothe identities' irreducible searches for control" (66), and are able to account for physical and social uncertainties (87). Indeed, "stories come from and become a medium for control efforts" (68). Because of stories, "a social network is a network of meanings" (67). "Culture and structure are not autonomous systems, they are intertwined and interdependent sets of social formations" (Godart/White 2010: 581). Moreover, stories are both objects and means in the study of markets. Stories are not only devices but also provide the *data* with which we can analyze both content and structures of markets.

With the linguistic, narrative, and cognitive turns sweeping through the social sciences since the 1980s, we have witnessed a surge of attention to the role of language, speech acts, words, rhetoric, metaphors, storytelling, narratives, stories, communication, texts, and discourse. Language, its role and usage, came to be understood as ubiquitous, important, and accessible for empirical investigation into meaning-making and actor construction.

Sociologists have always been working with words and narratives as data since "narratives are packed with sociological information, and a great deal of our empirical evidence is in narrative form" (Franzosi 1998: 517).[10] In their investigations of individuals and larger social structures, sociologists have always used interviews and documents to gain insights. By now, sociological practice includes data from sources such as newspaper articles, letters, autobiographies, directories, registers, speeches, contracts, manuals, reports, scientific articles and their abstracts, online tweets, and posted comments. Analysts use the data as a way to gather background information and to systematically interpret what is going on using different methodological approaches. Some reconstruct narratives; some interpret their meaning; others point to rhetorical figures or metaphors, or search for metanarratives or grand discourses. Still others classify and extract events, actions, or actors from the texts, count and quantify them, or establish relations between elements in those texts. The toolbox of social science methods is filled with diverse approaches to analyzing verbal and written texts (e.g., Czarniawska 2004; Elliott 2005; Franzosi 2010; Grimmer/Stewart 2013; Holstein/Gubrium 2012; Krippendorff 2012; Mayring 2015; Riessman 1993). Organizational scholars show that organizations can be studied through discourses or metaphors (e.g., Alvesson/Kärreman 2000; Putnam/Boys 2006), and, indeed, that organizations are discursively constructed (e.g., Fairhurst/Putnam 2004; Grant et al. 2004; Putnam/Fairhurst 2001; Putnam/Cooren 2004; Taylor/van Every 2000, 2011). Similarly, economists study the rhetoric of economics (e.g., McCloskey 1998).

Certainly, the field of research on language and communication in economic settings is very large and heterogeneous. The plurality of disciplinary, theoretical, and methodological traditions, pursuing diverse explanatory aims, using empirical materials of different scope and form, and placing different emphases on which elements of language and communications should be analyzed, has led to rather separate scholarly discussions. Nevertheless, these approaches share the very basic idea that linguistic, narrative, discursive, or semantic elements capture cultural meanings of actors acting, which, in some way to be specified, construct actors and their social situation, and which can be analyzed. I take this as the central idea for the ensuing discussion in which I bring different strands of research together in order to improve my analytical model and explanatory theory when studying stories both as devices for the emergence of markets and as data to study new social formations.

This first subsection introduces the conceptual and empirical usage of language and communication in current scholarship. It groups together approaches according to their relation of communication and actors. One stream of research focuses on the performative work of communication, creating and shaping actors. Another stream points to the co-constitutiveness of narratives and actors in interaction. A third stream of scholarship shifts the analytical focus to model social processes and structure using narratives, their words, and vocabularies as data. In delineating these different three streams on how the role of language and communication can be conceptualized, I provide a synthetic perspective and show the utility of my own sociologically informed take on the role of stories for markets and how to study them.

For now, I use "communication" as a general term to denote the exchange of information using language (written or in speech). A simple model of language holds that it is a mirror of social reality and transports meaning. The transfer of ideas from one actor to another is unproblematic since language is a medium. Communicating actors then offer a direct avenue into individual or collective cognition. Researchers are able to pick up this reflection on reality. However, sociological and organizational research has found that while language is very accessible for research purposes, its meaning is ambiguous, depending on interactions and context. Moreover, while research agrees on the constitutive potential of various types of communications for organizing a social formation, it disagrees on how exactly and with what outcomes communication participates in the organizing.

Research has shown that in particular social processes, communication is used to *perform* the social situation, or the actors involved: it affects the processes and "makes them true."[11]

One way communication is used performatively is by *storytelling*, either in personal stories (e.g., Boje 1991; Polletta 2009) or as a management tool to establish goals and visions (e.g., Brown et al. 2005; Denning 2005; Gabriel 2000). Such stories then also help in organizing: they manage knowledge (e.g., Schreyögg/Koch 2005), give reason and legitimize decisions (e.g., Lounsbury/Glynn 2001), convey and shape knowledge (e.g., Orr 1996), serve as a competitive organizational resource (Suddaby et al. 2010), and create communities within organizations (e.g., Bragd et al. 2008). In this sense, stories establish relations between different actors, actions, goals, and intentions; they do "relational work" (Tilly 2006: 70).

Other scholars emphasize the performative role of *discourses* from a Foucauldian perspective. Discourses in this view are the historically and culturally located systems of power and knowledge, which as collective practices, construct social order and meaning. As systems of statements, discourses exhibit formative rules that make the statements possible and shape how concepts can be used. Moreover, there is a dual relationship between power and discourse: discourse is an instrument of power, power affects discourse. Language has performative effects in that discursive practices construct the actors and the social order. Such works use a version of discourse analysis (e.g., Fairclough 1995, 2005, 2013; Keller 2013; van Dijk 2001; Wodak 2013).[12]

Economists studying *rhetoric* and *metaphors* of the economy have also long highlighted the persuasive and performative force of economic arguments for the economy using insights of literary theory (Klamer et al. 1988; McCloskey 1998; McCloskey/Klamer 1995). Economic models are metaphors, which work best in predictions, simulations, and counterfactuals (McCloskey 1995). Furthermore, "economics are saturated with narratives" (McCloskey 1990b: 9): economic stories explain something that happened and they are expressions of beliefs. In the long run, economic expertises may turn out to be self-fulfilling prophecies (McCloskey 1990a) and economic narratives may be driving economic events (Akerlof/Shiller 2009; Shiller 2019).

In sum, this stream of research understands communication as performative in the sense that it creates and shapes economic actors and situations. Here, communication serves as a tool to legitimize decisions, to tell of goals and visions, and to shape organizations, institutions, economies, or entire systems of knowledge. Not all communications may work according to the intentions of the speakers.

A second stream emphasizes the significance of narratives for *interpretation and meaning-making* among interacting actors. One focus is on *narratives*. It provides insights into how interpretation and meaning-making in the use of language and communication take place. Narrative in this view "is a kind of cognitive and cultural ether that permeates and energizes everything that goes on" (Pentland 1999: 712). Originating in literary theory, a narrative is a constructed account of a sequence of events, which consists of a plot and a story. Plots, as opposed to chronologies, require interpretation by an audience (White 1980). Narratives are thus a subcategory of the general idea of language, requiring emplotment *and* coherence in addition to a story, and communicate a mix of evaluation and description. Narratives are constructed after events have occurred; they have an end. By contrast, stories in this literary tradition are without emplotment, without beginnings, middles, and endings. What is more, narratives are not only individuals' cognitive structures used to explain how the world works, but they also take place in conversations between and across actors. Narratives exchanged between organizations show how meanings are constructed in interaction and how they are conveyed in narrative form to systematically constitute the organization (e.g., according to dramaturgical models, Czarniawska 1997, 1998).[13] Actors use "narrative embedding" to hold the ontological uncertainty about the future at bay: they interpret the context in which they must act based on past experience, thus developing a forward-looking "narrative logic," which they then follow (Lane/Maxfield 2005: 11). Because of narratives' temporal emplotment and possible interpretation thereof, organizational scholars of innovation have used the notion of narrative to understand organizing processes (e.g., Bartel/Garud 2009; Garud et al. 2014a; Garud et al. 2011).

One prominent approach to think about processes of meaning-making is by way of *sensemaking* (Weick 1995). Sensemaking presents an evolutionary model of organizing, which entails actors building narrative accounts, to organize their experience, to give meaning to events in order "create order and make retrospective sense of what occurs" (Weick 1993: 635). Moreover, actors make sense of events, actions, and objects in reference to an interpretive scheme or system of meaning. Sensemaking "is an issue of language, talk, and communication. Situations, organizations, and environments are talked into existence" (Weick et al. 2005: 409). In such narratives, actions are explained by intentions, and events are explained by causal relationships. Actors retroactively fit their experience to cognitive maps or frames, which classify, weave together different events, and suggest responses, thus giving it a "sense." Moreover, when new events occur, "narrative innovation" may be necessary and meaning and narratives get revised (Abolafia 2010: 356). Some scholars view organizations "first and foremost as communicative phenomena," since all "organizations are invoked and maintained in and through communicative practices" (Schoeneborn et al. 2014: 286). Given these pervasive communicative acts, sensemaking occurs interactively and collectively (Cornelissen/Clarke 2010).

Another set of scholars examine heterogeneous assemblages of devices to show how economic models create economic phenomena and shape economic behavior (e.g., Callon 1998a, 2007b; MacKenzie 2006; MacKenzie/Millo 2003; MacKenzie et al. 2007). Studying technical, material, and textual devices that incorporate economic models and theories, they show how these devices, rather than just beliefs or rhetoric, alter how things are done in the economy. One type of material device is the text—oral or written—which can for example contest, confirm, assure, and suggest in order to induce human actors to act (e.g., Cooren 2004; see also Callon et al. 1986). Such texts and documents "display a form of agency by doing things that humans alone could not do" (Cooren 2004: 244). Narratives in this sense are textual devices that provide opportunities for economic actors to judge and to calculate. When studying markets, they are part of a set of market devices, which are "the material and discursive assemblages that intervene in the construction of markets" (Muniesa et al. 2007: 2). Together with other devices and actors, narratives develop a force, by which they make "the economy itself as a communicative field and as an empirical fact" (Holmes 2009: 384). Analytically, these studies based in actor-network theory focus on several actors—nonhuman or human—and follow their narrative in interaction with other actors. However, narratives' role in meaning-making is not an explicit focus of ANT's analysis. Devices and assemblages are prime.

This second stream of research extends the performative view and considers the multiple interactions involved in communicating. It posits that communications are ongoing processes of local, situated interactions between diverse actors, by which all involved get co-constituted. Collective cognition and shared understandings get constantly produced and reproduced using language. Speakers cannot control what meaning others ascribe; rather, ambiguity and indeterminacy about meaning are expected. In effect, actors construct a social context through their narratives and their interpretation of what they and others do.

A third stream of research shifts the analytical focus to model social processes and structure using narratives, their words, and vocabularies as *data*. Whereas approaches in the discussions so far typically work with qualitative, interpretative methods in their analyses on the performative and co-constitutive role of communication for organizing and narratives for meaning-making, this stream uses formal structural methods to analyze patterns of narratives, words, or vocabularies in order to get at the cultural meanings of relations, processes, and structure *embedded in* organizations and their environments. Such cultural analyses thus show the "deep structures" of how social structures, such as organizations, are organized. In extension to new institutionalists' focus on the significance of symbols and cultural processes in the study of organizations and their environments (DiMaggio/Powell 1991), this stream of research uses formal methods and quantifies qualitative textual data to reduce "the complexity of meaning to simpler structural principles" (Mohr 1998: 345). This stream speaks directly to the relational approach as outlined in the introduction, when it folds together the measurement of meaning and social structures—since both are mutually constitutive—within the same research design using formal mathematical techniques (see also Basov et al. 2020; Mohr et al. 2020; Mohr/Rawlings 2015; Mützel/Breiger 2020; Roth/Basov 2020).

Methodologically, in order to "measure meaning structures" (Mohr 1998), analysts need to assess relations between cultural elements "by looking at how actors, organizations, or institutions make practical use of the cultural distinctions being investigated" (353). Then analysts need to decide on what basis these identified elements can be compared, i.e., how to "lump and split" the elements (Zerubavel 1996). In a third step, the related elements are analyzed with tools that allow insights into structural properties.

One such tool is sequence analysis, which allows one to model narratives of events in form of sequences, since "social reality happens in sequences of actions located within constraining or enabling structures" (Abbott 1992: 428). A meaning of an event is not only conditional on its temporal and social context, but also on its position in a sequence of other interrelated events (Abbott 2007b). This structural analysis of events holds a "processual understanding in which outcomes, actors, and relations are all endogenous" (Abbott 2007b: 19). For instance, narratives of events may be careers or life courses (e.g., Abbott/

Hrycak 1990; Aisenbrey/Fasang 2017; Blair-Loy 1999; Han/Moen 1999; Stovel et al. 1996) or disputes (e.g., Mützel 2002). Another such tool is network analysis working with words, sentences, or concepts as nodes. Considering narrative sequences of life stories as structural and representable networks that facilitate systematic comparison and interpretation of qualitative data, the approach of *narrative networks* (Stovel/Koski-Karell 2015) has modeled and explained the social processes involved in identity formation (Bearman/Stovel 2000; Padgett 2018; Smith 2007) and the meaning of events (Bearman et al. 1999). The approach breaks down a narrative into elements, i.e., network nodes, that are causally or temporally connected to one another, i.e., network arcs. Network analysis can then detect general patterns in the narratives. Other works have been concerned with larger discursive patternings of particular fields using network analysis, for example, changes in the structure of moral discourse (Mohr 1994), in institutional discourses (Mohr/Lee 2000), and strategies (McLean 2007).

As the discussion shows, the field is heterogeneous in theoretical and methodological approaches, empirical focus, and explanatory goals. My analytical aim was to group works together according to how they understand the relation between cultural meanings and the situations, actors, or social structures they study. I have identified three different streams. These approaches have in common that they focus on narratives, or parts thereof, that have been collected *ex post*, with endings, in particularly defined social settings, i.e., a field or an organization, in which it is clear who participates in the communications. While some approaches stress the negotiation of meaning, understandings of some collectively shared meaning prevail. The perspective I advance in this book expands these approaches to encompass a cultural analysis of the emergence of a market, which traces the processes of emergence from multiple perspectives using a plurality of stories, and with multiple methods in order to detect meaning-making processes. These meaning-making processes necessarily involve negotiation of meaning and take place in conversations across actors.

I use the term *story* to denote a general, undogmatic notion of the narrated, from small texts to big narratives. Stories are an integral part of interaction and communication practices. They are what is told—in texts or speech—capturing both a reference to social time *and* content. They are mostly fragmented, because the ending of a story may not be determined yet. They do not have to be causally or temporally emplotted, though they may be. They may also be reports or analyses. They may entail factual elements as well as imaginations and expectations about what the future will hold. The perspective I take here is that each text is a story. An individual text can also be part of a story or at least attempt to be part of a story. Another text may connect to something stated much earlier, inferring already established meanings. Stories are part of conversations across different actors; recursively, they also shape each actor's identity. Stories in this sense are both *data* for analysis as well as *devices* in the formation of a market.

Focus of the empirical case is on a market that is just beginning to emerge, namely, the market for innovative breast cancer therapeutics in the late 1980s. My argument is that in order to understand how actions are coordinated when it is unclear who participates and who will profit, we need to view markets as networks of sociocognitive and sociomaterial actors and focus on the stories of the market. Market participants are involved in telling of and listening to stories, as they try to establish whom to evaluate and what to interpret and whom to watch as equivalent such situations of uncertainty. Therefore, what is being told may be very fragmented, unfinished, and evolving. An *ex post* perspective is difficult to uphold. Moreover, while some of what is being told concerns direct interpretation of the past to obtain guidance for the present and future, this is not the case for all stories. Some stories are factual reports, asking for interpretation and evaluation of third-party experts. Other stories are expectations or stories of the future.

So far, the role of public stories of third actors, fragmented narratives, stories of the future and their projectivity, as well as the role of ambiguity and how categories help to temporarily settle ambiguity have been absent from the three research streams discussed. For a cultural analysis of an emergent social formation, however, they are most relevant.

Public stories, e.g., in news or business reports, provide not only factual news, but also interpretations and evaluations. As "brokers," "promissory organizations," or "intermediaries" (Martin 2015; Pollock/Williams 2010, 2016), news, industry, and finance organizations watch and comment on what they see going on in markets. Media reports, such as journalists' stories, analysts' reports, or press statements, provide associations between companies and products by mentioning them in the same publication. These third parties are thus involved in sensemaking and providing evaluation of what is comparable to what else: media evaluators are "helping producers and consumers alike make sense of new offerings and reactions to them" (Kennedy 2005: 209) and thus of what can be cognitively connected across organizational boundaries.

These connections can be examined using terms and their co-occurrence across newspaper articles (Kennedy 2005). Also, by mentioning other firms or products in self-issued press releases, firms establish a cognitive connection between themselves and others in public stories. For the study of emerging markets such "cognitive embedding" (Kennedy 2008) by co-mentions is of particular relevance since it "provides a way to observe the associations that are necessary to construct new market categories" (273).

Of similar importance is the fragmented, incomplete nature of stories. Organizational scholars have indicated that there are stories that are too polyphonic, fragmented, nonlinear, unconstructed, and incoherent to be already analyzed as narratives with a plot and an ending. They point to analyze "fragments of communication, conversation, or text that construct identities and interests in time and space" (Vaara/Tienari 2011: 373), which in their unstructured way may be a pre-narrative of the future (Boje 2001). They are projections, based on the present, of what the future will bring.

Such a bet on the future is also at the heart of the "sociology of expectations," which empirically focuses on innovation processes of technologies (e.g., Borup et al. 2006; Brown/Michael 2003; Brown et al. 2000; Hedgecoe/Martin 2003). Expectations here are understood as a "state of looking forward" (Borup et al. 2006: 286), as exhibited in "imaginings, expectations and visions" (Borup et al. 2006: 285). Such expectations capture what has been called *projectivity* (Emirbayer/Mische 1998: 971):

> the imaginative generation by actors of possible future trajectories of action, in which received structures of thought and action may be creatively reconfigured in relation to actors' hopes, fears, and desires for the future.

They are "composed of creative as well as willful foresight" (Mische 2009: 697). These expectations may not become true; they may need to be adjusted as time passes due to cognitive limitation and contextual changes, such as new actors and new information. Yet, while being oriented toward the future, they have a real influence on decisions and actions in the present as they motivate, organize, and coordinate action also oriented toward the future (Beckert 2016; Tavory/Eliasoph 2013). These expectations of possible future trajectories of action are narrated in form of *stories*. "Expectations, and stories about the future in general, reduce essential contingency in a non-deterministic sense, by providing blueprints that can be used in action" (van Lente/Rip 1998: 217). As such, stories are "forceful fiction" (van Lente 1993) and provide a "narrative infrastructure"

(Deuten/Rip 2000), i.e., the "rails along which multi-actor and multi-level processes gain thrust and direction" (74), constitutive of innovation processes.[14]

Recent scholarship on such future-oriented stories that shape the present and coordinate action have called them "projective stories" (Garud et al. 2014b), "future projections" (Mische 2009), or "fictional expectations" (Beckert 2013, 2016, 2019; Beckert/Bronk 2018).[15] While emphasis and empirical field of each term is slightly different, they all conceptualize the role of imaginaries of the future as well as the link between cognition and action from a pragmatic understanding of action. The premise is the fundamental uncertainty and unpredictability of the future. Past or present empirical evidence can be rationally calculated and extrapolated into the future. However, actors are faced with cognitive limitations and contextual shifts that severely reduce the ability to make accurate predictions.

Stories, as discussed above, provide reason and rationale, interpretation, and evaluation of what is going on. *Stories of the future*, with their imaginations and expectations about what the future holds, provide orientation "*despite* the uncertainty inherent in the situation" (Beckert 2013: 222). They do so by offering "convincing" accounts of the future, which are comprehensible and plausible, thereby momentarily "overlooking" uncertainty (222). As Beckert summarizes his argument (226), quoting Searle (1975, 328):

> Fictional expectations represent future events *as if* they were true, making actors capable of acting purposefully with reference to an uncertain future, even though this future is indeed unknown, unpredictable, and therefore only *pretended* in the fiction expectations. . . . [E]xpectations are, under conditions of uncertainty, "pretended representations of a future state of affairs."

Because of this ability to temporarily suspend uncertainty, stories of the future are fundamental for economic actions. To be sure, regardless of whether stories of the future correctly predict the future, they nevertheless influence action in the present. Stories of the future are performative not in the sense that they help create an economy according to economic models, but rather in the sense that they orient and motivate others to act because their offered future seems believable and desirable. As prominent examples from financial markets show, such stories can shape expectations of others *causing* the events that they predicted (Beckert 2016).[16]

Empirically, we find such future-oriented stories in situations in which something new is "emerging" such as in a new technological field (e.g., Selin

2007), when it is unclear what the new may be a case of (Kennedy 2008), who the legitimate actors are (Lounsbury/Glynn 2001), which products and projects may be realized (Kaplan/Vakili 2014), and which strategy should be followed albeit temporarily (Kaplan/Orlikowski 2013). Entrepreneurial projects start in stories of the future, which, in their promises and imaginaries of the future, help to maintain or regain entrepreneurs' legitimacy (Garud et al. 2014b).[17]

Two aspects of such *stories of the future*, as I will call them, are worth noting further. One is their *relational* character: they "are 'peopled' with others whose actions and reactions are seen as intertwined" (Mische 2009: 701). Stories are the ties that connect actions and actors. Moreover, when heard they may influence third actors' action in the present or in the future. Another aspect speaks to the *temporal patternings* of stories of the future. Their imaginations may reach into short-, middle-, or long-term futures (699). Moreover, when viewed over time, different moments of narrating and acting exist. For example, "emerging technologies rely on promising stories to garner support in the early stages" (Selin 2008: 1884). Early stories of scientific developments may be more hypothetical, as factual knowledge may be limited. Such stories are prone to "narrative innovations," i.e., revisions of what the future may hold. They are provisional narrations, which may change over time as actors interpret and make sense (Bartel/Garud 2009).

Stories of the future not only "overlook" uncertainty; they also may provide for productive ambiguity. An emergent market is already a highly uncertain setting, in which it is unclear who is a participant and what the products are. Moreover, it is also a setting with extreme ambiguity: it is unclear what a story may mean (Santos/Eisenhardt 2009: 644), potentially leading to multiple and competing interpretations.[18] This *ambiguity* is a resource for economic activity (Esposito 2013; Stark 2009) and is indeed foundational for capitalist dynamics (Beckert 2013, 2016). As indicated earlier, ambiguity leads to misunderstandings and dissonance (Stark 2009). In turn, dissonance provides for disruptions, frictions, and surprises necessary and productive for newness (de Vaan et al. 2015; Hutter 2011; Hutter/Stark 2015).

Ambiguity, then, is not only part of an encountered situation or encountered actor that needs to be coped with. Rather, entrepreneurs can induce ambiguity in their stories, thereby turning it into a strategic resource (Jarzabkowski et al. 2010; Sillince et al. 2012). Entrepreneurs can exploit ambiguity about multiple evaluative frames and thus interpretation of situations and actors, for example when they are engaged in "discursive pragmatism" (Stark

2009: 108). Rather than declaring a problem as solved, actors may agree to provisional settlements, for instance in order to meet deadlines: the provisional settlements offer momentary cognitive clusters of agreement of what the situation is about. However, misunderstandings and multiple frames of evaluation simmer beneath such temporary fixtures. Differences are not "ironed out," leaving room for reinterpretation, multivocality, and potentially permanent innovation (Stark 2009).

In order to coordinate amid uncertainty and ambiguity, actors use stories to temporarily settle on *categories* about what the situation is about. Categories seemingly reduce cognitive complexities of social reality and provide a "conceptual system" (Rosa et al. 1999). Indeed, works on sociocognitive processes of classification and categorization in markets have highlighted that these processes are essential for the existence of markets (e.g., Aspers 2009; Beckert/Aspers 2011; Breiger 2005; Negro et al. 2010). Studies on these processes have shown that they serve several purposes, in particular in the market making: they define boundaries of who and what count as legitimate and meaningful (e.g., Kennedy 2008; Lamont/Molnar 2002); they generate shared understandings about producers' identities and their products (e.g., Rosa et al. 1999; Zhao 2005) and thus act as institutionalized reference of what can be evaluated with what evaluative measure of comparability (e.g., Espeland/Stevens 1998; Zuckerman 1999). Categories thus help to make sense of market elements. Categorization and classification processes are rife with social processes of valuation and evaluation (Durand et al. 2017; Lamont 2012).

A growing number of studies examine how involved market actors established shared meanings of new market categories. For example, Khaire and Wadhwani (2010) show how "actors construct new meanings of a category and then translate them into referents that enable valuation within that category" (1282). Others point to the role of underlying cultural codes for the existence of new categories (e.g., Weber et al. 2008). Such studies on categorization processes view stability of categories as necessary for the examined markets to emerge, because stability allows for shared meaning and anchors comparison.

However, in emerging markets, meanings are neither collective, nor stable, nor shared. In emerging markets, meaningful categories only come to stick over time, when a first label gets adopted and is used by different actors (Grodal et al. 2015). Labels may stem from attempts to self-categorize (e.g., Vergne 2012; Zhao et al. 2013), though such labeling is often contested (Cattani et al. 2017; Granqvist et al. 2013; Slavich et al. 2020; Vergne/Swain 2017) as it needs to resonate with audiences or other intermediaries (Hsu et al. 2012; Navis/Glynn 2010; Pontikes 2012). Moreover, labels may serve as supportive vocabularies in communicating a new category (e.g., Hsu/Grodal 2021; Loewenstein et al. 2012; Nigam/Ocasio 2010; Ocasio/Joseph 2005). Especially new combinations of words and concepts help to signal a new category, while also allowing for common ground to understand the meaning of that new category (e.g., Grodal et al. 2015).

"Category emergence" and "category creation" are two possible avenues category formation can take (Durand/Khaire 2017). In the process of "category creation," actors—in mostly cognitive moves—redefine and reinterpret preexisting attributes and features to elaborate how a new category is different from and normatively superior to an existing category (e.g., Rao et al. 2003). Instead, processes of category emergence involve a labeling *after* a material innovation has been identified by involved actors in the market. In category emergence, the category acquires its legitimacy through the explication of the meaning of the category—most notably from new companies or third-party intermediaries in discursive contestations (Grodal/Kahl 2017; Slavich et al. 2020). Intermediaries help to interpret and evaluate a new category, provide meaning, and facilitate an agreement between different actors in the nascent market on what their activities are about. By drawing attention to the contested meaning-making processes of a diversity of actors involved underlying the emergence of a new category, the categorization of innovation can be seen. In addition to the new category, new market actors and products are outcomes of category emergence.

Markets from Stories

The proposed analytical model and explanatory theory on markets from stories synthesizes the building blocks this chapter has introduced.

In the first part of the chapter, the discussion of prominent perspectives in economic sociology focused on what constitutes markets. It identified direct relational ties and their meaning for the actors, observation of competitors, sociotechnical arrangements, and orders of worth. It also pointed to the need to interrelate different aspects so that dynamics of markets and a diversity of actors can be grappled with. Moreover, existing perspectives typically apply to markets that already exist or provide an *ex post* account of a market came to be constituted. The chapter then discussed two recent works that specifically address the emergence of newness in general, and the emergence of markets more specifically. Discussions of these frameworks yielded two results: structuralists models alone do not suffice to explain how actors position themselves vis-à-vis one another, or how they evaluate ideas or concepts; interpretation and evaluation are necessary to allow comparison and establish equivalence.

I propose a combination of both approaches with further extensions. Such combination yields a model of multiple network domains that need to be studied over time, with a focus on reflexive cognition, interpretation, and evaluation. In addition, the model I propose accounts for a crucial element that allows for interpretation and evaluation of markets: stories.

The chapter's second section explored the role of stories in research involving economic actors. The discussion identified several patterns: one stream of research focuses on the performative work of communication, creating and shaping actors; another group points to the co-constitutiveness of communications and actors. A third group of scholarship shifts focus to use narratives and language as data for analyses of the cultural. The perspective I advance in this book expands these perspectives to a cultural analysis of the emergence of a market, which traces the processes of emergence from multiple perspectives using a plurality of stories, and with multiple methods in order to detect meaning-making processes.

Like the discussion of market constitution, research on stories has most often focused on already existing social formations, such as organizations or markets. Stories on emergent processes, however, involve third-party evaluators, are fragmented, concern some uncertain future, and are full of ambiguities. As a result, this book proposes analyzing the making of categories as a temporal phenomenon from which actors switch in and out and which may be stories of the future. Such a perspective considers that categories in an emerging market are not fixed entities to which all can agree. Since it is unclear what will count as worthy (Stark 2009), prior classification schemes using "prototypes" (Glynn/Navis 2013) or taxonomies (Bowker/Star 1999) against which a product can be compared are difficult to apply. Categories themselves emerge.

This study advances work on the categorizing of innovation when it views categorization processes as entangled processes across heterogeneous actors and their stories. Together and interrelatedly, firms' and third-actor public stories are involved in contested meaning-making processes on categories. Out of entangled stories, which bundle together meanings about what something is about, categories emerge in particular situations and moments. For some actors these categories may be sharply drawn; for others their boundaries may be fuzzy. Some categories may stick; others categorization projects may fail.

When studying the emergence of a market from a processual perspective, the making of categories has its own dynamic. Categories "emerge and fall out of use" and can be analyzed as such without inquiring into the legitimacy or illegitimacy of a category (Kennedy/Fiss 2013: 1138). A category begins as a label and may experience rivalry from other labels. There may be competing stories about what constitutes a label; there may be alternative labels. A category may enroll its own support, just like Callon's notion of translation. Only over time and when the label gets adopted and used by different actors, it may become a meaningful category (Grodal et al. 2015; Grodal/Kahl 2017). Thus, categories are also more than just cultural constructions: categories too become actors around which others can rally and mobilize. They are generative for further actions.

Thus the book advances a relational and processual perspective on markets as networks of sociocognitive and sociomaterial actors, in which actors tell stories. *Markets from stories* is a model of multiple network domains that need to be studied over time, with a focus on meaning-making explored through analyses of stories.

Rivals not only watch each other's actions, as White's original market model posits, but they also tell stories and listen to each other's stories. Their actions and their stories about these actions serve as signals to all actors in the market, about their prior market situation, their current situation, and their future trajectory, and also relationally affect all other involved actors.

Stories, then, are the ties that connect actors and actions over time and help to constitute each actor's identity (White 1992, 2008). Stories of the future help to motivate, orientate, and coordinate actions. Stories are thus primary for contextual, situational meaning-making. Stories help to establish an interpretation of the perceived market structure and help, albeit momentarily, to stabilize market profiles, suspend competition, reduce uncertainty, and mobilize financial resources. In the long run, a market structure emerges. As White expands in his work on business discourses, markets are created, used, and reproduced by participating actors in a network of stories as the principal medium of this social construction (2000). Thus stories are at once about the conveyance of *information* and *interpretation* as well as the speaker's structural position.

Stories here are a means to inquire into the relational constitution over time of actors, objects, and processes, providing descriptions and evaluation. This perspective follows a basic tenet of both network analysis and pragmatist actor-network theory: there is no *a priori* ascription of who the powerful actors are.

As an analyst, I am agnostic; I follow all sorts of actors in making associations. Positions of power get established in the processes of making connections, and may only be of temporary relevance. Much like in ethnomethodology, texts serve as the actually observable data. In particular, I use public texts to get at the public construction of markets. I view stories as devices and data to get at agency and process. They help to model processes of emergence across multiple network domains over time.

Stories serve as the cognitive connections while "assemblages of heterogeneous actors" (Çalışkan/Callon 2010), including theories of organizing, tools, materials, and humans, are needed to create a market. Indeed, money, organizations, people, tools, tests, and other nonhuman actors are needed as well for a market to emerge. Moreover, it is not one story that is told, but rather many stories of many actors vying to participate in defining what counts in a market. In order to "follow the actors" (Latour 2005) and to tell a story of marketization processes (Callon 2015), the analysis will have to take into account a plurality of stories and perspectives and a plurality of connections concomitantly criss-crossing each actor: in other words, an actor is shaped by more than its egocentric relations and by more than a single story.

Additionally, in an emergent market it is unclear how stories can be evaluated since it is unclear what will count as worthy. Multiple evaluative principles are at play. Stories of the future make promises and pretend to know what the future will hold. Also, there may be multiple and competing interpretations of these stories of the future. Actors compete for "their story" and interpretation to be of worth to others. Stories may imitate other stories (we are comparable to others); they may also anti-imitate and dis-associate themselves from already existing stories (we are like no one else) (Abbott 2007b; White 2002). In these evaluation processes, meaning is locally and temporarily shared. Coordination among actors amid such ambiguity occurs via stories and their temporary, situational entanglements in categories.

Such a perspective, the book suggests, allows a focus on the evaluative and narrative dimensions in emergent processes. In effect, the book offers a cultural analysis that shows crucial elements and mechanisms of market emergence, including the categorization of innovation. The next chapter introduces the empirical field and data for such an analysis.

2 Breast Cancer Therapies and Innovation

FOR CENTURIES, BREAST CANCER has been the target of treatments in attempts to prolong the life of those who suffer from it. In the early twentieth century, several invasive therapies, including radiotherapy and chemotherapy, were developed, which until the 1970s remained standard treatment options. Their therapeutic mechanism rests on the destruction of growing cells, which include cancerous but also healthy cells. The quest to find a "magic bullet" for curing breast cancer without harming healthy tissue, proposed as early as 1900, continued to motivate research. In the 1980s, this idea became the "holy grail" of clinical research on cancer. Several research approaches developed by biotech companies, such as interferon or monoclonal antibodies, initially promised to be such a magic bullet. Exploring the field of breast cancer research in these earlier periods is essential to comprehending the context in which innovative breast cancer therapy research and the initial formation of a market took place.

A look at the early biotechnology industry shows how these new enterprises at the intersection of research and business translated ideas of basic research into commercial products using technologies of genetic engineering. It also shows a first cycle of biotech's innovation-driven expectations and disappointments. At the end of the 1980s, biotechs were unable to meet outrageously high self-set and media-promoted expectations to develop such a magic bullet. The high hopes for a nontoxic treatment of cancer were stifled when developed drugs failed crucial clinical trials; patients and doctors were disappointed, and investors withdrew their support.

This chapter lays important groundwork in introducing the empirical field, the field of commercial breast cancer research. It describes in general terms what breast cancer is and offers a brief history of the treatment of breast cancer, including molecular oncological research and highlighting general therapeutic approaches,[1] until the late 1980s. The last section of this chapter presents the field of biotechnology and cancer research at the end of the 1980s.

Breast cancer is currently the most frequently diagnosed cancer and the leading cause of cancer deaths for women worldwide. In numbers: in 2020, globally, an estimated 685,000 women died of breast cancer and 2.3 million women, i.e., 1 in 4 cancer cases in women, were newly diagnosed breast cancer patients (Sung et al. 2021).[2] Because of improved diagnostics and detection, but also reproductive, hormonal, and lifestyle risk factors, the numbers of new cases (*incidences*) have increased worldwide over the past decade. Incidences and mortality vary across the globe: whereas *mortality rates* have been declining in highly developed countries since the late 1990s, breast cancer mortality has steeply increased in sub-Saharan African countries and Oceania (see Figure 2.1 for a region-specific look at incidence and mortality rates). For the year 2030, the IARC/WHO (2020b) predicts 2.74 million new breast cancer patients.

Cancer is a group of diseases that is characterized by the uncontrolled growth and spread of abnormal cells. Abnormal cells can be caused by external (environmental) or internal factors. Environmental causes are chemicals, radiation, and viruses. Internal causes include hormones, immune conditions, and inherited mutations. All these factors may work together to initiate the disease, which may not manifest itself until ten or more years after original exposure.

When the abnormal cells become malignant, the cells invade and destroy normal tissue. In the beginning, cancer cells usually remain at their original site and the cancer is localized. The cancer cells may then invade distant or neighboring organs or tissue. This can occur either by direct extension of growth, or by metastasis, whereby the cells become detached and are carried through the blood or lymph system to other parts of the body. Metastasis may be regional, or it may spread throughout the body. When this occurs, the condition is referred to as advanced cancer.

In most general terms, breast cancer is a malignant tumor that involves abnormal cell growth in the milk-producing glands of the breast or in the ducts that deliver milk to the nipples, with the potential to spread to other parts of the body. As opposed to leukemias, breast cancers are solid tumors.

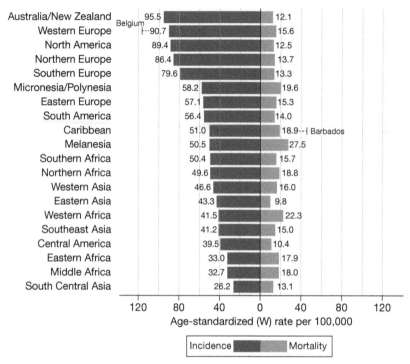

Figure 2.1 Region-Specific Incidence and Mortality Age-Standardized Rates for Female Breast Cancer in 2020. Source: IARC/WHO 2020a

Brief History of the Treatment of Breast Cancer

As a disease, breast cancer has been recognized and has been the target of attempted treatments for thousands of years, as records from Ancient Egypt, China, India, and Persia dating from several centuries BC attest.[3] Hippocrates of Cos (approx. 460–370 BC) first used the words "carcinos" and "carcinoma," from the Greek term "karkinos," referring to a crab. In his writing, he was comparing the long, distended veins radiating from a breast tumor to the limbs of a crab. Galen of Pergamon (approx. 130–200 AD) elaborated on Hippocrates's system and wrote about cancer of different organs. Greek and Roman physicians developed the theory of humoralism on the makeup and functioning of the human body. According to it, a body is filled with four basic substances (or humors): blood, yellow bile, black bile, and phlegm. They are in balance when a body is healthy. According to this humoralist understanding, cancer is caused by a bodily imbalance of an accumulated excess of "black bile" at the afflicted body site.[4] Although treating patients with herbs and special diets, to restore the bodily balance, Galen also held, that once somebody was diagnosed with cancer, there was no definite cure.

In the sixteenth century, scientists began to develop a better understanding of the body. Renaissance physicians gained new knowledge on surgery and the role of lymph nodes in cancer. Andreas Vesalius of Padua (1514–1564) challenged humoralism by failing to confirm the existence of black bile. Some surgeons began the practice of removing the axillary lymph nodes when removing cancerous breasts. In the seventeenth century, insights on the recirculation of the blood and the existence of lymphatic vessels questioned the existence of black bile. Physicians began to combine surgery with lymphatic drainage. As antecedents of what later was to be known as "radical mastectomy" in the nineteenth century, European physicians developed guidelines for total mastectomies, removing lymph and ribs, and developed new surgical instruments, which allowed for quicker amputation of the breast. Surgery was seen as an effective treatment of localized breast cancer, but many believed that once cancer was in the lymphatic system, it was inoperable and fatal. At the end of the eighteenth century, humoralism was effectively replaced by a theory on the role of the lymphatic system and an idea of a local origin of cancer.

During the nineteenth century, medicine in general and the knowledge on cancer were truly transformed. Cell theory showed that cells are the basic building blocks of the human body and that they are created by cell division. Cell theory, along with microscopic analysis of the body, replaced lymph theory and others. It led physicians to focus on the local growth of cancer. Important developments were systemic analyses of hereditary cancer based on genealogical records and improvements in hygiene during surgery and afterward. Moreover, "radical mastectomy," the *en bloc* removal of the entire breast, regional lymph nodes, as well as all the muscle tissue of the breast, began to be performed. In comparison to previous types of treatments, radical mastectomy brought about impressive results of survival. This type of surgery, which was developed in the late 1880s, influenced treatment strategies up until the 1960s. Indeed, until the development of chemotherapeutic treatments and a better understanding of cancer leading to breast-conserving surgery, or lumpectomy, radical mastectomy was the most successful and most applied treatment.

Modified, simple mastectomy, without the removal of the muscle tissue of the breast, is still performed today as a treatment to eradicate cancer from the female body for late-stage breast cancers. For women who carry certain mutations of the genes BRCA1 or BRCA2, most commonly an inherited mutation that accounts for approximately 5–10% of all breast cancer cases in women (e.g., Campeau et al. 2008) and that increases the lifetime risk of developing breast or ovarian cancer, bilateral prophylactic mastectomy is currently a standard treatment to reduce that risk of breast cancer. Developments in anesthetics and antibiotics have helped the performance of such surgeries. Breast-conserving lumpectomy or even minimally invasive surgery is now a medical option in early-stage breast cancer or after chemotherapeutic treatment. Knowledge on cell biology and thus the identification of tumorous cells has helped to limit the incisions to the necessary circumference.

Radiotherapy, developed after the invention of X-rays in the 1890s, was only rarely used at the beginning of the twentieth century, as its force was too destructive, doing more harm than help. The usage of radiation as a therapeutic treatment in breast cancer really started in the 1950s, when surgery shifted from radical to simple mastectomies and when radium was replaced with cobalt-60, emitting less destructive γ-rays. Radiotherapy helps to destroy tumorous cells and is typically used as adjuvant, i.e., additional, therapy after surgery. It has become a standard treatment after lumpectomy or breast-conserving incision. Radiotherapy's typical side effects are soreness of the skin, swelling of the breast, and fatigue.

The idea of chemotherapy, understood as the controlled use of chemicals to treat a disease, was first developed around 1900. However, it was not until after World War II that the era of chemotherapy really began (e.g., Chabner/Roberts 2005; DeVita/Chu 2008). Pathbreaking studies proving the concept of chemotherapy were conducted during the 1940s. They showed that alkylating agents, such as compounds related to mustard nitrogen, as well as folic acid analogues, reduced tumor sizes. During the 1950s, first clinical trials on breast cancer patients were conducted that showed that systemic therapy with chemotherapeutic, cytotoxic agents sends tumors into remission. During the 1960s and 1970s much research on chemotherapeutics was conducted, searching for the best combination of toxins with the least side effects. Adjuvant chemotherapy following breast cancer surgery, which improved survival, became the standard regimen during the 1970s.[5] Beginning in the 1980s, chemotherapy was also applied as neoadjuvant therapy, shrinking breast tumor sizes *before* surgery and preventing the growth of metastases. Common side effects of chemotherapy are fatigue, nausea, and diarrhea, all bodily reactions to the fact that chemotherapeutic toxins attack all cells of the body. Also, high-dose chemotherapies with tremendous, debilitating side effects for cancer patients, as well as many new compounds and new combinations, were tested during the 1980s. By the 1990s, the initial optimism about advances in combinatorial chemotherapy as a cure-all for cancer had waned. While cytotoxic agents proved highly destructive for rare cancers, such effectiveness did not occur in more common cancers, such as breast cancer. Patients developed chemotherapeutic resistances, and breast cancers recurred at high rates.

Researchers gained new insights into hormonal dependence of breast cancers with advances in endocrinology and cell biology. While estrogen therapy proved effective as early as the 1940s in some postmenopausal breast cancer patients, research on the estrogen receptor during the 1960s gave new rise to hormonal therapeutics. Tamoxifen, initially developed as a contraceptive, turned out to function as an antagonist of the estrogen receptor. It blocks estrogen from binding to its receptor on tumor cells and thus prevents cancer cell growth. Because of this mechanism, it only works for patients with estrogen receptor positive (ER+) breast cancer. The Food and Drug Administration (FDA)[6] first approved tamoxifen in 1977 for advanced breast cancer patients. Since then, its indication has been expanded: tamoxifen is prescribed as an adjuvant therapy following surgery or chemotherapy, to prevent breast cancer recurrence. It is also given to women who are at high risk given their ER+ status to prevent the occurrence of breast cancer in the first place. It is taken as a daily pill over the course of several years. Other hormonal therapies focus on aromatase inhibitors, which block the activity of the enzyme aromatase, necessary in estrogen production.

In summary, at the end of the 1980s, treatments for breast cancer included surgery, radiation, chemotherapy, and hormonal therapy. Separately or in combination, these options proved to be effective treatments and often enough brought about remission of the disease. Nevertheless, and despite all advances in treatments, mortality rates due to breast cancer remained high. Problems encountered in particular are gradual resistances to existing treatments, unresponsiveness to existing treatments, unpredictable metastatic growth, and several side effects. Surgery is not always able to eliminate all cancer cells. Chemotherapy is a toxic and nonspecific treatment that aggressively destroys all cells in the process, thus causing nausea and fatigue for the patient. Radiation also destroys cells locally, leads to fatigue, and is not a reliable technique. Hormonal therapy only works for ER+ breast cancer patients.

As early as 1900, chemical researchers had proposed the idea that foreign microbes, and similarly cancer cells, might possess biochemically unique binding sites whose blockade by a "magic bullet" could result in the targeted destruction of the foreign/cancer cell without harming healthy tissue around it (Strebhardt/Ullrich 2008).[7] In the 1980s, this idea of a "magic bullet" blocking certain biochemical mechanisms became the "holy grail" of clinical research on cancer, just as the search for a factor residing in the cell-producing tumors was the holy grail of basic research (Fujimura 1996: 64).

In 1976, researchers proposed that normal human cellular genes, so called proto-oncogenes, could be activated to become oncogenes and thus to cause cancer (Stehelin et al. 1976). This proto-oncogene theory vastly impacted cancer research in the years to come.[8] It explained how radiation, tobacco, or other irritations could initiate cancer by mutating and thus activating proto-oncogenes within a cell. It also explained why the same kind of cancer can be found in people with different exposure to irritation: both smokers and nonsmokers have the same proto-oncogenes in their cells; however, smokers develop cancer at a higher rate because carcinogens in tobacco increase the mutation rate of these genes. But the proto-oncogene theory not only provided a theory for the prevention of cancer. It also shaped technologies and strategies of cancer research.

Indeed, in the early 1980s, biologists and cancer researchers began to incorporate new genetic technologies, in particular recombinant DNA (rDNA) technologies, into their laboratories; they soon became "the tools of the trade" (Fujimura 1996). By the mid-1980s, basic cancer research was extensively molecular, with proto-oncogene research as its most prominent line of research. Several proto-oncogenes were isolated directly from cancer cells, e.g., *ras*, *myc*, or *src*, located inside the cancerous cell, as well as *neu*, located at the cell membrane with a fragment hanging outside of it. Soon tumor suppressor genes, so called anti-oncogenes, were also identified.

The conceptual framework for carcinogenesis, i.e., the formation of cancer cells from normal cells, then became that (Mukherjee 2010: 381)

> the cancer cell was a broken, deranged machine. Oncogenes were its jammed accelerators and inactivated tumor suppressors its missing brakes.

Three new ideas for attacking cancer emerged from those insights in the late 1980s (406): cancer therapeutics could attack 1) the genes that accelerated the growth of cancer, while saving healthy cells; 2) the pathways of cellular signaling, where proto-oncogenes and tumor suppressor genes are located and where cancer cells are activated; 3) cancer cells' growth of blood vessels and their resistance to death signals. Oncogene research had thus pointed at basic mechanisms in the search of a "magic bullet": something needed to be switched on or off, and this something could be found inside, outside, or at the interstices of tumorous cells. These three ideas will be important in guiding cancer research for the next decades and help to structure the field of biotechnology and cancer research.

The Field of Biotechnology and Cancer Research at the End of the 1980s

Developments in cancer research and the evolution of the field of biotechnology are entangled with each other. In the late 1970s and early 1980s, "amphibious" entrepreneurs (Powell/Sandholtz 2012a) founded biotech companies in research-intensive areas, geographically located close to university lab settings, such as the Bay Area, San Diego, and Boston in the United States, Cambridge and Oxford in the United Kingdom, e.g., Cetus (Emeryville, CA; 1972; IPO 1981), Genentech (South San Francisco, CA; 1976; IPO 1980), Amgen (Thousand Oaks, CA; 1980; IPO 1983), Hybritech (La Jolla, CA; 1978; IPO 1981), Biogen (Geneva, Switzerland, Cambridge, MA; 1978; IPO 1983), Genzyme (Boston, MA; 1981; IPO 1986), Chiron (Emeryville, CA; 1981; IPO 1983), British Biotechnology, later British Biotech (Oxford, UK; 1986; IPO 1992).

These entrepreneurs (Padgett/Powell 2012a: 375)

> simultaneously occupied leadership positions in academic labs and entrepreneurial ventures.... The result was the creation of a new form—a science-based commercial entity, which emerged from overlapping networks of science, finance, and commerce.

These research-driven companies, using new knowledge on genetic engineering and molecular biology, thus worked as alternatives to both basic research at university labs and big pharmaceutical companies, while at the same time pursuing profitable interests. In the United States, also venture capital, private and institutional investors, helped to start up these biotech firms. These small firms then entered collaborative, contractual relationships with university labs and pharmaceutical companies to develop, market, or license products. In Europe, different institutional frameworks (Casper/Matraves 2003), collaborative ties, and research foci (Owen-Smith et al. 2002) have led to a different, less innovative and competitive path than in the United States (Allansdottir et al. 2002). Yet here, too, new companies were founded, pursuing profitable interests (Hodgson 1990). Rich studies on the structure and logics of the field of biotech have shown the role of interorganizational ties between funding agencies, investors, and universities for the development of biotech innovations (e.g., Casper/Murray 2005; Ebers/Powell 2007; Galambos/Sturchio 1996, 1998; Henderson et al. 1999; Jong 2006; Kaplan/Murray 2010; Kenney 1986a; Kenney/Florida 2000; Pisano 2006; Powell 1996, 1998; Powell/Brantley 1992; Powell et al. 1996; Powell/Owen-Smith 1998; Powell/Sandholtz 2012b; Powell et al. 2005; Teitelman 1989; White et al. 2004).

Looking at the emergence of the field retrospectively, these new companies crossed boundaries and effected transpositions (Powell/Sandholtz 2012b: 384). While

> the people who built the commercial field of biotechnology lacked any formal blueprint of constructing a [biotechnology firm], each carried tacit blueprints from the domains they knew well.

Together they came to create a new organizational form: a model of a science-based company that was unlike the traditional vertically organized corporate hierarchy of pharmaceutical companies. At the same time, however, this new field and the new organizational forms had to sustain themselves amid different evaluative criteria. A diversity of forms generated "divergent standards and multiple kinds of rules, resulting in competing criteria for gauging success" (Powell et al. 2012: 438).

Genentech provides an illustrative example on how biotechs translated the ideas of basic research into commercial products using technologies of genetic engineering. This will point to evaluative toggling that the new companies in the new field were facing. It will also be a substantive reference for later discussions.

Soon after its founding, the biotech company Genentech began to focus on the commercial exploitation of rDNA technologies.[9] With those technologies, genes could be transferred from one organism to another; genes could also be spliced and recombined into new genes, thus creating proteins not found in nature. Many complex drugs are proteins, such as insulin, which used to be produced by extracting animal innards. Using the recombinant DNA techniques and bacterial cells as "factories," Genentech was able to produce recombinant human insulin in cooperation with Eli Lilly's French facilities. In 1982, Genentech's human insulin was the first recombinant drug to receive FDA approval (Hughes 2006).[10]

Like other biotech start-ups at that time, Genentech participated in the media-hyped rush to genetically develop interferon, an antiviral agent, which in the late 1970s and early 1980s was seen as a "non-toxic wonder drug" against all cancers. Teaming up with big pharma Schering-Plough (Kenilworth, NJ), Biogen was the first biotech to produce interferon alpha (Pieters 2005). Since 1980, Hoffmann–La Roche (Basel, Switzerland) was among the collaborators of Genentech's interferon research.

Genentech also worked on the development of monoclonal antibodies (mAbs), which are large-molecule proteins that bind to specific substances, e.g., to tumor-specific antigens that are located at the surface of a cancer cell.[11] In theory, mAbs would then "selectively seek out those cells and destroy them" (Klausner 1986: 185), also carrying chemotherapeutic or radiotherapeutic agents. Moreover, Genentech researchers worked on the identification of proto-oncogenes as receptors of proteins, for example epidermal growth factor (EGF) (Ullrich et al. 1984; see also Hughes 2006, 2011).

Genentech researchers, like others involved in basic cancer research, discovered the *neu* oncogene in the mid-1980s and called it HER2/neu.[12] Whereas HER2/neu is a proto-oncogene in normal cells that helps them grow, they understood that it becomes an oncogene, a hyperactive protein in a human cell, when a cell has too many copies of this gene. However, it was unclear for which cancer it was hyperactive and also how it might be inactivated. By testing cell samples of cancer patients, collaborating lab researchers found that HER2 was overamplified in some breast cancers but not in all. In addition, HER2 positive (HER2+) breast cancers were found to be particularly aggressive (Slamon et al. 1987).[13] Genentech researchers believed that HER2 was a target to which monoclonal antibodies could attach and deactivate the growth of cancer cells. In animal tests, HER2+ breast cancer tumors disappeared after treatment with the antibody (Bazell 1998: 42). Researchers apparently had found a switch to turn off the growth of cancer cells.

Yet in 1988 Genentech management decided to stop this line of research, pointing to difficulties with the commercialization of antibodies and with anticancer products more generally, as Genentech's interferon product had failed in clinical trials. Several scientists of the HER2 project left Genentech at the end of the 1980s to pursue other projects.[14]

In addition to Genentech's human insulin (licensed to Eli Lilly), the FDA approved other drugs based on recombinant DNA technologies. Genentech received approval for its human growth hormone (1985), interferon alpha-2a

(1986) for hairy cell leukemia (licensed to Hoffmann–La Roche), and the blood clod dissolver tPA (called Activase, 1987). In 1989, Amgen received FDA approval for its engineered erythropoietin (EPO). In addition to further interferons, interleukins also were in the pipeline as anticancer treatments. Interleukins are a group of proteins that occur naturally in human bodies to regulate the white blood cells and thus induce immune responses (Klausner 1988). In 1989, Cetus received approval for selling its product interleukin-2 as a treatment for kidney cancer in several Western European countries. At that time, the company expected to receive FDA approval of Proleukin for other cancers as well, including breast cancer: "progress was encouraging in the use of the product against other kinds of cancers of the lung, breast, ovary and bladder" (Marsh 1989).[15] As Kaplan and Murray (2010: 125) summarize the developments of the time: "All signs pointed to a burgeoning industry that would yield medically efficacious products to save lives and deliver high investor returns."

Yet none of these approvals was uncontroversial. Legal and clinical problems ensued (e.g., Gladwell 1988). And although sales reached dimensions unheard-of, as in the case of Activase and EPO, stock prices fell. As Powell and Brantley note for the case of Genentech's blood clot dissolver Activase (Powell/Brantley 1996: 244):

> Genentech was a victim of its own extraordinarily high goals, which were uncritically embraced by Wall Street and which in turn further fueled the biotechnology community's expectations. . . . Genentech needed to recoup both its $200 million development costs and the expenses of building the organizational capability to market it. It also needed revenues to fund its new product development.

It turned out that at the end of the 1980s, the field of biotechnology disappointed investors, often by failing outrageously high self-set expectations, which were also created and intensified by media and analysts' reports (see also Klausner 1988). Management of these amphibious enterprise-meets-research labs proved more difficult than expected. In particular, evidence on the basis of clinical trials proved more difficult to obtain than expected.

Moreover, patients were disappointed. Interferon, for example, once hailed as the "miracle drug for cancer," only showed a response rate of 10–15% in clinical trials (except for hairy cell leukemia). It was "no magic bullet" after all (Sun 1981). Success turned out "to be a highly ambiguous term" (Pieters 1998: 1232):

> Overall response rates ("efficacy") resulting from clinical studies under

controlled circumstances might look promising, even when it remained unclear what this actually meant for individual chances of success and how well a treatment might perform in everyday clinical practice ("effectiveness").

The "initial 'Genentech' model of biotechnology [had] failed as a commercial project" (Kaplan/Murray 2010: 125). Drugs failed crucial clinical trials, patients and doctors were disappointed, and investors withdrew their support. The high hopes for a nontoxic treatment of cancer were stifled. Early in 1990, Genentech executives decided they "could no longer afford the huge development and marketing costs required by their products" (Powell/Brantley 1996: 246). The Swiss pharmaceutical company Roche Holding, owner of Hoffman-La Roche, purchased 60% of Genentech in February 1990, with an option to buy the remainder over several years (Business Wire 1990; Ratner 1990; Russell 1990).

This is the setting we enter at the end of the 1980s. It is a setting where new research approaches were promising, yet a path to approved products was unclear, with disillusioned investors and disappointed patients. The involved actors of the newly formed biotech industry were reconfiguring their conceptualizations of the fields of biotech and cancer therapeutics.

3

A Market of Expectations

NARRATIVES OF SCIENTIFIC PROGRESS often emphasize singular individuals who make a pivotal "discovery," and point to their successes. Certainly, the field of cancer research has produced many groundbreaking treatments—*Her-2: The Making of Herceptin* (Bazell 1998) and the take on the *Magic Cancer Bullet*, Gleevec (Vasella/Slater 2003), provide details into the developments of such treatments.[1] Yet from the relational sociological perspective, we know that many more actors are involved in developing innovations and that there are many more failed innovations than successful ones. At the same time, focus has often been on a single innovation, even when tracing its evolution from different perspectives over time (e.g., Latour 1996).

Thus the challenge becomes how to capture the developments of the entire field of breast cancer therapeutics' emerging market. In this chapter, I do this by identifying general research strategies in the field of breast cancer therapeutics from the late 1980s until the late 1990s. Based on these research strategies and using qualitative data, I closely follow the developments in the early years, before any product existed, up until first approval and its repercussions. This chapter thus traces how early participants in the North American and European field of "innovative breast cancer therapeutics" are trying to make sense of what they are about and with whom they are competing, by examining their stories. The analysis highlights ambiguities about newly developed molecules as well as biotechs' research strategies. Ultimately this highlights how cognitively interrelated actors come to collaboratively construct a market of expectations, which, at least temporarily, turns out to be a profitable strategy.

In order to follow the developments as a market is emerging, the analysis starts from the beginning of these scientific discoveries.[2] As a first step, I reconstructed the scientific discussion on innovative breast cancer therapy with the help of the journals *Nature* and *Science* and the oncological literature from the late 1980s and until the late 1990s and matched the different biochemical research approaches to more general research strategies. I checked this match with experts in the field and the oncological literature. I use these albeit temporary research strategies as heuristics to analyze the field of biotech involved in innovative cancer research.[3]

The idea of "innovative breast cancer therapy" as it was discussed during the 1990s exhibits different research strategies. One way to conceptualize these strategies, and thus picking up the formative ideas of oncogene research, is by focusing on the relationship of cancer cells to other cells. Oncogene research had pointed at basic mechanisms: something needed to be switched on or off, and this something could be found inside, outside, or at the interstices of tumorous cells.

One strategy followed by cancer research concentrates on the *extracellular environment*. Research hypotheses then focus on three mechanisms: antiangiogenesis, or the stoppage of blood supply and thus of all nutrients to the tumorous cell; vascular targeting, which prevents the buildup of blood supply in the first place; or cancer growth inhibitors, which prevent communication between cancer cells. Chemically, these mechanisms to stop the cancerous behavior can be carried out by either small-molecule compounds or monoclonal antibodies (mAbs), which are large molecules. In contrast to small molecules, mAbs are unable to penetrate a cell membrane. Matrix metalloproteinase (MMP) inhibitors block key enzymes that are responsible for angiogenesis and tumor growth (Brower 1999).

Another strategy centers on the *interstices between the extra and intracellular environment*. Such a strategy focuses on the development of immunotherapeutic agents that would enable an artificial activation of the immune system, in turn immunizing patients against the antigens that are expressed in cancer cells with the goal of eradicating these cells (e.g., Huber/Wölfel 2004; Mellman et al. 2011).

A third strategy focuses on the *intracellular environment*. Two competing working hypotheses relate to this strategy: tumor suppressors directly attack tumorous cells and block the cell's cycle (e.g., tumor suppressor protein p53) (e.g., Campisi 2005); or molecule-delivered suicide genes work to self-destruct tumorous cells (apoptosis) (e.g., Garber 2005; Melnikova/Golden 2004).

A fourth strategy uses a *combinatorial and conjugate approach*: it combines elements of the previous three strategies with chemotherapy or hormone replacement therapy.

From the data set comprising press statements, newspaper articles, and reports from industry analysts, I identified all companies claiming to work on a molecular cancer therapy, which could be effective in the case of breast cancer, or to which analysts ascribe such a potential.[4] In 1991, four companies were part of that group—Biomira, Bristol-Myers Squibb, Genentech, and NeoRx—of which two were biotechs. Such an explorative procedure allowed me to take into account companies that were once active, though at a later time merged or dissolved or whose product turned out to work for a different indication. Adhering to the standards of qualitative text analysis (Miles/Huberman 1994), I read and manually coded the 845 texts for biotech companies, their research strategies, evaluation of research strategies, and particular events taking place in multiple rounds to find relations and patterns.[5]

The following analysis reconstructs the market emergence of innovative breast cancer therapy research as it evolved between sixteen North American and European companies at the end of the 1980s until 1998. It considers all biotech companies that at some point made a substantive claim in the available sources that they were developing molecular, innovative breast cancer treatment. The analysis focuses on reports of biotech companies themselves, their financial and industry analysts, the scientific community, and, in turn, the ensuing reaction of different market participants.[6] Communication about different research strategies serves as a heuristic to understand biotech companies' positioning in the emerging market (Penan 1996). Stories about the chosen research strategy and its expectable results function as signals, which competitors interpret and use to orient their own actions in next steps. Excerpts from these stories highlight the attempts of positioning and showcase the role of expectations and hope. Moreover, taken together these public stories collaboratively construct the market as its meaning and its worth is negotiated between heterogeneous networks of actors.

Positionings, Findings, and Expectations

In the 1980s, researchers at Genentech successfully develop[7] molecular antibodies that prevent vascular growth, and specifically the growth of cancer cells in animal models. By 1988 they have identified "an oncogene, a form of cancer that specifically activated that oncogene, and a drug that specifically targeted it" (Mukherjee 2010: 417). Genentech nevertheless abandons its cancer research because of high costs and slight chances of expected success. With limited funds, some researchers at Genentech and the University of California, Los Angeles, continue to work on the HER2 project and in 1990 they have produced a humanized HER2 antibody, ready to be used in clinical trials, called *trastuzumab* (419).

The Canadian biotech company Biomira (Edmonton, AB; 1985; IPO 1988) follows a different research strategy. Since the late 1980s, Biomira has been working on antigens that stimulate the body's immune system into producing antibodies, which resist cancer. In 1989 animal tests of Biomira's cancer vaccine that fights existing tumors first prove successful (Pharmaceutical Business News 1989).

In April 1991, Genentech conducts phase I of clinical trials of HER2, at UCLA with twenty women to test the drug's safety, and declares (Business Wire 1991):

> The preclinical data are very strong. We're hopeful that the human therapeutic effect of the HER-2 antibody's interference in tumor growth will be confirmed.

Comments from industry analysts are somewhat more skeptical (Pharmaceutical Business News 1991):

> The use of monoclonals derived from mice have had disappointing results in human trials, but the Genentech team have fused a mouse monoclonal to a human antibody called immunoglobulin-G, with the hope of increasing the immune system's chance against cancer.

Other biotechs are also participating in these developments on breast cancer therapies, each with particular research strategies. The biotech Sugen (Redwood City, CA; 1991; IPO 1994) is founded on the scientific insights into a crucial mechanism to block cell signaling; it specializes in small molecules that block tyrosine kinase receptors outside of tumorous cells (Business Wire

1992). IDEC Pharmaceuticals (Mountain View and La Jolla, CA; 1986; IPO 1991) works on immunotherapeutic approaches for lymphoma and melanoma, with the intention to start preclinical trials on solid tumors in 1991 (PR Newswire 1990). Others work on mAbs that carry chemotherapeutics or radiotherapeutics to be delivered directly to the cancerous cell, for example NeoRx (Seattle, WA; 1986; IPO 1988) and ImmunoGen (IPO 1989; Cambridge, MA; 1981; IPO 1989).

Then, in January 1992, Biomira's stock prices skyrocket (Dow 1992):

> [T]he collective wisdom of investors is valuing Biomira at almost $400 million. Four hundred million dollars! Four hundred million dollars for a company that has never made a profit . . . ? For a company that admits it expects to incur further losses in "the foreseeable future"? For a company whose sexiest products (cancer treatments) have yet to be produced or marketed commercially? Can Biomira be worth $400 million?

Industry analysts are quick to link this stock movement to a research report written by financial analysts, who prematurely hinted at an FDA approval of Biomira's vaccine, while clinical testing had not even begun (Jorgensen 1992). By May 1992, stock prices had dropped by 60%. In July 1992, Biomira successfully completes phase I (testing toxicity) of clinical trials (PR Newswire 1992):

> Having treated 24 patients with our Theratope formulation we believe the Phase I studies show that the product is safe and stimulates their immune systems to respond against the cancer antigen.

In September 1992, Biomira begins phase II trials (testing safety and efficacy) in Canada, also for the treatment of breast cancer. At this time, Medarex (Princeton, NJ; 1987; IPO 1991) and Chiron (Emeryville, CA; 1981; IPO 1983) move into phase I of clinical trials with so-called bispecific antibodies that bind to both HER2 *and* immune cells, with the idea that the immune cells would destroy the cancer cells (Business Wire 1993; Weiner et al. 1995).

An industry analyst at that time points to the inherent difficulty of establishing evaluative measures for these biotech companies (Gilbert 1992):

> Valuation will continue to be the single most difficult issue for the biotechnology group, with uncertainty about valuation adding to price volatility throughout 1992. Most companies lack traditional valuation parameters such as sales and earnings and, instead, trade on future expectations.

In fall 1993, Biomira starts phase II clinical trials in the United States for late-stage breast cancer patients. A reporter notes (Cowley 1993):

> No one is touting it as a cure. Theratope is not designed to prevent cancer in healthy people but to prevent relapses in cancer patients who are in remission. In preliminary studies, it has added only months to patients' lives, but it embodies a fresh approach to mobilizing the body's defences. . . . If immunotherapy fulfills its promise, some cancers may become as beatable as a bad flu.

Also in the fall of 1993 and after successful phase I trials (Chase 1992), Genentech begins phase II clinical trials for its HER2 humanized monoclonal antibody for treating metastatic breast cancer that is resistant to chemotherapy, enrolling forty-three patients (Business Wire 1994b).

Meanwhile in Europe, especially in the United Kingdom, biotech companies follow the leads of the North American biotech companies and begin to develop molecular cancer therapies.[8] Since 1992, British Biotech (Oxford, UK; 1986; IPO 1992) is working on a novel enzyme inhibitor, which would "limit tumor spread, particularly in highly invasive cancers such as breast, lung and ovary" (Daily Mail 1992). By February 1993, British Biotech has started phase I of clinical trials.

The research strategy behind British Biotech's enzyme inhibitor focuses on attacking the extracellular environment of tumorous cells, in this case by hindering the formation of new capillaries (an inhibitor of matrix metalloproteinase, MMP). As the first product of this particular biochemical mechanism, the compound called Batimastat shows success in phase I. It was tested on an uncommon cancer in fifteen patients, yet, as British Biotech management points out: "'It could have implications for other kinds of cancer. That's the chief hope.'" Indeed, "the drug is potentially a 'blockbuster'—meaning sales could run into hundreds of millions of pounds" (Skeel 1994). Investment analysts are enthusiastic: "if any UK company is going to make the big-time, it is British Biotechnology" (Investors Chronicle 1994a). Commentators also acknowledge what this means for cancer patients (Christie 1994): the new drug tested in Edinburgh

> raises the long-term hope that cancer could become a condition that people can live with, safely controlled from spreading through a regular drug treatment regime.

While the supervising medical doctors point out that "a great deal more work is needed to understand the process," they also "hope the drug will lead to a new way of attacking primary cancers and containing them before they can spread" (Hawkes 1994). After the encouraging news of the trials, British Biotech successfully floats more shares to raise funding for further drug development (Green 1994).

In May 1994, British Biotech begins phase II of clinical trials with Batimastat in the United Kingdom (Financial Times 1994). In June 1994, British Biotech announces in a press statement that preliminary results of current clinical trials show that a marker of disease progression is decreasing in the tested patients. Moreover, it is developing a new compound, i.e., an advanced Batimastat, to be taken orally, which is expected to enter clinical trials before the end of 1994 (Business Wire 1994a):

> If successful in trials, it is anticipated that it could have applications in as wide a range of cancers as batimastat, including breast, lung, colorectal, prostate and many others. However, the oral drug would be targeted at patients who are at an earlier stage of disease progression.

Financial analysts evaluate British Biotech and its compounds very positively. While "valuing a biotech share is largely guesswork in the early development phases," the recommendation under the heading "British Biotech—Patient hope" is nevertheless to buy its shares: "Batimastat targets a market worth at least UK£ 400 million each year" and but if "lead drugs progress through their next set of large-scale clinical trials the shares will soar. Bio is high risk, but buy ahead of that" (Investors Chronicle 1994b). Others are similarly enthusiastic in their evaluation (Lumsden 1994):

> If Batimastat proves helpful in treating ovarian cancer, it is also likely to be useful against a wide range of other cancers—and this is what has driven the share price to its current peak. With 6 million cancer sufferers in the US and Europe, the group could have a blockbuster.

In November 1994, phase III clinical tests start for Batimastat, which three hundred patients in Europe will test. Analysts estimate Batimastat's potential sales until 2000 to range in the US$ 250 million and its probability of success at 35% (Mccrone 1994)—and with such estimates, stock prices rise. British Biotech also announces that its improved product, now called Marimastat, which works with the same mechanisms and can be taken simply as a pill, has begun phase I of clinical human trials.

In January 1995, the biotech company Chiroscience (Cambridge, UK; 1992; IPO 1994) announces that following the same mechanism it has developed a more effective product than British Biotech (Daily Mail 1995). British Biotech, however, reports strong side effects from Batimastat's phase III trials. Initially it points to problems in production due to large volumes and assures that the production process will be altered and one of two trials only temporarily suspended (Business Wire 1995). This announcement sends British Biotech's shares "plunging" (Evening Standard 1995). The founder of Chiroscience uses the opportunity of the weakened competition to promote his company in a press release (Pharmaceutical Business News 1995):

> We have the best pipeline of any emerging pharmaceutical company with a research programme that will give us the blockbuster drugs of the future.

However, British Biotech counters promptly and starts with phase II for Marimastat in May 1995. Positive preliminary test results for Marimastat and the promise of further human trials let stock prices and the analysts' expectations rise once more (Alexander 1995):

> British Biotech, though it is not claiming anything as headlinish as a "breakthrough" in cancer treatment, made a presentation to City analysts about its drug Marimastat. It left some of them clawing the air. The shares rocketed from 1043p to 1548p. A breathless Ian White of Flemings [investment banking] hastily penned a manuscript note for clients. It is the "opportunity of the decade," he scribbled. The last time he wrote what he calls a four-star note was just ahead of Glaxo's launch of Zantac, which proved to be the world's most profitable drug.

Given the estimated annual sales of more than 600 million British pound for the finished product, London's financial world is agitated (Stevenson/Wilkie 1995):

> The City has whipped itself into a frenzy in the belief that it can cash in on the holy grail of medical research—a cure for cancer. In their scramble not to miss out on the billions to be made from a cure, investors are brushing aside warnings from the medical establishment that it is too early to tell if a breakthrough has been made.

Nature's *Biotechnology* publication is also enthusiastic about MMP inhibitors as potential cancer drugs (Hodgson 1995).

Meanwhile, in North America, biotechs also report on their clinical trials. In March 1995, Biomira reports from their phase II trials and its stock prices soar (Biotechnology Business News 1995a):

> More than 50 per cent of patients in breast and ovarian adenocarcinoma groups are still alive. The data indicates the therapeutic vaccine is "as good or possibly better than chemotherapy, without the toxicity associated with chemotherapeutic agents."

Also having positive results from phase II of clinical trials (Baselga et al. 1996), Genentech then is able to begin the crucial and most difficult phase III of clinical trials in fall of 1995, and thanks in part to a compassionate use program and eventually a change in regulations that got rid of the requirement to test against placebos, is able to recruit enough women for the trials.[9] In October 1995, Sugen discloses that it is working on a HER2 inhibitor compound, working on a different receptor than Genentech was (Biotechnology Business News 1995b). In March 1996, scientists at the Memorial Sloan Kettering Cancer Center in New York find "first clinical evidence that a new Genentech Inc. treatment can help in advanced cases of breast cancer" (Hall 1996). Also, Biomira's trial results are positive, prolonging life of breast cancer patients (Lanthier 1996).

In April 1996, Chiroscience yet again declares that animal tests show that "our product is better than Marimastat" (Extel Examiner 1996). Subsequently, financial analysts recommend buying Chiroscience stocks, whose value increase rapidly. Also in April 1996, the US firm Agouron Pharmaceuticals (La Jolla, CA; 1984; IPO 1989) presents its preclinical studies, which show that working on the same mechanism, it too has a better product than Batimastat in the pipeline (PR Newswire 1996).

In May 1996, phase III tests for Marimastat begin. As an industry analyst reports, financial analysts react enthusiastically to this development: "We think it's going to be a major cancer product" (Davidson 1996). Stocks rise very high and British Biotech is for a short while a company without a product but worth US $ 2.45 billion.[10]

However, the scientific community criticizes the presented results from the clinical trials: they are based on falsely collected data, which measure natural variation and not, as claimed, a stop of tumor growth (Anonymous 1996). Financial analysts, in turn, blame British Biotech with deceit, and although British Biotech goes on to defend itself—"We have been careful what we have said.

We never claimed Marimastat is a cure for cancer, but we hope it will be a useful treatment" (Blackstone 1996)—stock prices fall. In November 1996, British Biotech stops the development of Batimastat.

At about the same time, three other British biotechs try to find their position in the emerging market for innovative breast cancer therapeutics. Cyclacel (founded in 1996 with capital from one of the co-owners of Chiroscience, Dundee, UK) promises to modify and suppress cancer growth with a "potentially revolutionary approach," the protein p53 (The Times 1996). Oxford Biomedica (Oxford, UK; 1995; IPO 1996) announces yet another strategy to tackle tumorous growth: the destruction of tumor cells with the help of gene therapy. Antisoma (London, UK; 1988; IPO 1997) works on an immunotherapeutic therapeutic. In the fall of 1996, the German company Medigene (Martinsried, Germany; 1994; IPO 2000) enters this field and works on immunotherapy. Marimastat's clinical trials are taking longer than planned and the results are not satisfactory. Financial analysts waver between high expectations and pessimism.

In May 1997, Biomira announces a partnership with US biotech Chiron, an expert on vaccines, to further develop and test Theratope (Lanthier 1997). News media are excited about the future prospects of cancer therapies (Fisher 1997):

> The biotechnology industry is poised to deliver a host of cancer drugs in the next two to three years. Some of them promise to extend lives—or at least make treatment more bearable than chemotherapy and radiation. . . . If the biotechnology companies are as successful as they expect to be, the end of the decade should see many more new therapies for cancers of the breast, prostate, lung, colon, liver, ovary, pancreas and kidney, among others. Perhaps the biggest gain in the new therapies is that unlike chemotherapy drugs they have few and very mild side effects. Because they are so well tolerated, the drugs can be prescribed in large doses that are more likely to have an effect, and they can be taken for many years.

In a November 1997 press release, Genentech[11] announces successful preliminary results from phase III clinical trials (Business Wire 1997a):

> The anticancer activity of the investigational Herceptin monoclonal antibody may result in both slowing the progression of the cancer and increasing the percentage of women who experience tumor shrinkage. Although not a cure, these

encouraging results in women with metastatic breast cancer suggest that better control of the disease may be possible.

Genentech announces that it will enter the "oncology market" and attempts to stay in it with a "pipeline of innovative biotherapeutic agents" (Business Wire 1997b).[12] Meanwhile, Biomira and Chiron delay phase III trials for Theratope, in order to first test an improved version. Investors react "by driving the stock to its lowest level since 1995" (Bell 1997).

Chiroscience and Agouron enter phase II of clinical trials with their products; contrary to British Biotech, they partner up early on with pharmaceutical companies (Bristol-Myers Squibb and Roche, respectively). In April 1998, industry analysts view Chiroscience as more successful than British Biotech (Manley 1998), since Chiroscience's cancer therapy offers fewer side effects and better usability. The pressure to deliver better results from clinical trials shakes up British Biotech: the company's management leaves over disputes about publication of trial results, and there are legal investigations regarding the accuracy of press statements and insider trading (Massod 1998). As a result, British Biotech has lost its credibility and stocks nosedive completely (Frankfurter Allgemeine Zeitung 1998).[13]

In May 1998, Genentech applies to the FDA for approval of its breast cancer treatment Herceptin (*trastuzumab*) (Business Wire 1998b). The FDA grants its approval in September 1998 (Business Wire 1998a). In that press release, Genentech first thanks the nine hundred women who volunteered in the clinical trials. It then points out:

> Today heralds a new era in breast cancer with a new weapon that targets an underlying genetic defect that causes cancer.

Genentech's CEO adds:

> Our employees feel a strong sense of accomplishment with the approval of this new therapy, one that can truly make a difference in the lives of those it helps.

Politicians, too, praise Herceptin's positive results, as a reporter notes (Hall 1998):

> "We now have a new weapon in our fight against breast cancer," said Health and Human Services Secretary Donna Shalala. "For certain women with advanced disease, this new product can mean new hope."

With the sale of Herceptin's marketing rights outside of the United States to Roche, Genentech's earnings nearly double in fall 1998 (The New York Times 1998). With Herceptin, which shows fewer side effects and "durable objective responses" (Cobleigh et al. 1999), the innovative breast cancer therapeutics market has its first approved product.

Whereas a first phase of market emergence closes with that approval, expectations and hopes for new products applicable to more patients, and more profits, continue to propel all participants further. In October 1998, phase III trials of Theratope begin, about which clinical experts say (Hope 1998):

> Our early data shows it is not going to eradicate the disease but, for women who have undergone chemotherapy, it will act as an adjunct to hopefully keep the disease at bay and prolong patients' lives.

Stock prices climb with the beginning of the trials[14]—as do patients' hopes and expectations during that time. A newspaper story reports how "even the whisper of an effective cancer treatment can make desperate patients beg to become guinea pigs in research projects." Patients and their family members approach cancer researchers for possible treatment (Taylor 1998):

> A steady stream of new cancer therapies are emerging from the research labs, and every time they make headlines or pop up on the Internet, they have the power to launch cancer patients and their families on an emotional rollercoaster ride. And so the stories of hope and expectations continue.

Emerging Market of Innovative Breast Cancer Therapeutics

These narrative exchanges between biotech companies, industry and financial analysts, as well as members of the scientific community have shown the *entanglement* of developments in the field of breast cancer research. Interacting through their public stories, biotech companies tell other market actors where they position themselves and others in the market. The stories told by economic actors provide information on how they assess their situation and their context and how they act accordingly. Such stories then create a competitive social structure in which the involved actors position themselves vis-à-vis others and forge their own identities. This relational positioning influences not only the strategies of the actors directly involved but also the possibilities of others for obtaining R&D resources. As a result of the narrative establishment of positions, competition can be suspended, uncertainty temporarily reduced, and financial resources mobilized. As all biotechs and their analysts work with the biochemical knowledge and the financial resources available at that time, a *cognitive interdependence* between companies, strategies, molecules, and compounds becomes evident.

In their search for a "magic bullet against cancer," biotech companies pursue different research strategies in order to find the most successful therapy against tumor cells. Their choice of research hypothesis depends on various factors. For instance, the available financial and personnel resources may make some approaches unviable; companies may experience normative and cognitive lock-ins (Grabher 1993); the "imprinting" given when the firm was founded (Stinchcombe 1965), which is closely linked to the educational history of the scientific founders and the rest of the personnel, may influence the choice.

In this first phase, the companies involved pursued two research strategies in particular: anti-angiogenesis and immunotherapy. With an eye to the doings of Genentech, a successful pioneer, other biotechs worked on their own cancer treatments. Yet they also made use of the expectations that Genentech's research on anti-angiogenesis had generated. As the previous section has shown, British Biotech and Chiroscience directly contested each other for the same market niche, by both working on the same mechanism. Nonetheless, the story about the MMP-inhibitor approach allowed both companies to albeit temporarily build up their profiles, thus not only contesting but also confirming each other. Media attention around British Biotech's test results led stock prices to rise and fueled further expectations of economic profits and medical breakthroughs. Both companies exploited the positive outlook, as voiced in analysts' stories of the future, in order to secure financial resources.

Some of the stories told in this phase of market emergence were debunked by other actors as lies and replaced by financial analysts and the scientific community with new stories. Some biotech companies, such as Chiroscience, explicitly employed an imitation strategy (Tarde 1962) when they presented themselves in their stories as using a similar biochemical mechanism to successful competitors, but were still able to benefit by distinguishing themselves clearly as separate enterprises. Other companies, such as Antisoma and Biomira, which invested in immunotherapeutics, tried to find their own niche by using a strategy of anti-imitation. Moreover, clinical trials outcomes put the stories of the future to a test. In the event of negative results, new actors quickly gathered to tell their new stories—they communicated new expectations, they achieved positive reactions from the analysts, and they thus secured financial resources for themselves.

The qualitative inquiry of the first few years of the emerging social formation shows that actors primarily exchange *expectations about the future* in their stories. These stories speak of hopes and promises, point to expected outcomes in clinical tests as well as in financial valuation, and project a future with a cure for cancer. Prior to the existence of any molecular breast cancer product as such, biotechs are thus trading stories on expectations, projections, and imaginations on what the future holds. These stories provide orientation for the involved actors "*despite* the uncertainty inherent in the situation" (Beckert 2013: 222) and thus despite the ambiguity of how to evaluate these new molecules, because the stories offer plausible—and in many cases, much desired by not only the companies and investors, but also patients and the broader public—accounts of the future. Successful results from a phase II clinical trial get extrapolated into the future to also hold for a phase III; financial valuation and patients' expectations follow suit. Share prices reflect future expectations, and setbacks in research results damage such expectations. Stories of the future tie clinical results of the present to the scientific and thus also financial performance in the future. All actors telling stories therefore shape expectations about the future as a part of the present. In this way, they distance themselves from the uncertain future and can compete in coordination with each other for the most promising prospect. If stories change as a result of the arrival of new actors or new test results, then the structure of the market also changes.

But it is not only biotechs that tell stories to instigate interest for particular research strategies and thus to secure funding. In a similar narrative move, journalists and financial analysts voice their bets of the future as they must remain reputable and have the best and most valuable insights. They too are facing the ambiguities of newness and need to cope with the many unknown unknowns on how to interpret and evaluate biotechs' research reports. Hooking on to stories about the future full of expectations and promises helps all involved cope with uncertainties of the situation while it concomitantly strengthens their cognitive interdependence. Together and with their entangled stories, these actors are collaboratively creating a *market of expectations*.

Companies, analysts, and investors were betting their future on different research strategies with limited knowledge on the biochemical mechanisms, yet with great hopes on the future. Moreover, as biotech companies were convincing investors to provide financial support with their research findings and promising results, they were also trying to find out what they were about. The analysis of the early years of the breast cancer therapeutics markets thus also reveals that *heterogeneous criteria of worth* were at play. Biotechs evaluated their research results as worthy for science, their investors, and for patients— and thus also raised expectations regarding scientific success, possible profits, and finding a cure. Analysts evaluated companies because of their results and thus too participated in the construction of expectations into future treatments and investment opportunities. Breast cancer patients saw the worth of new molecules in a successful treatment. They were hopeful to have new and better treatment options soon.

The successful approval of Herceptin in 1998 as the first molecular product for the treatment of breast cancer changes the focus of the field overall. With approval, the *market of expectations without a product* becomes a *market with a product* to compare against, yet it remains full of expectations, promises, and stories of the future as now even more companies and investors join in.

4

Making Sense of the Market

WITH THE FDA APPROVAL of Herceptin in 1998, biotech companies working on breast cancer research had a reference to which they could compare themselves, their molecules, and their compounds. The fields of science, commerce, and finance similarly had a comparative reference. For the field of cancer research, this approval signaled the fundamental conceptual change that less toxic and less invasive cancer treatment was possible. It also suggested that "cancer will become understandable in terms of a small number of underlying principles" (Hanahan/Weinberg 2000: 57). Yet many questions about the underlying biochemical mechanisms and also further potential treatments remained unanswered. Once Herceptin and the HER2/neu mechanism was approved, several questions of interpretation and evaluation became central: What about other research strategies now that Herceptin potentially applies to 25% of all breast cancer patients? Was Herceptin and its biochemical mechanism an outlier or indeed the first mover of a new path of treatment?

While analysts of the cancer market were unsure how to evaluate the role of molecular cancer therapeutics during the 1990s and, in turn, how to categorize these new types of treatments, with the approval of Herceptin and other mAb-based cancer therapeutics, *a new treatment category* began to be needed. In this chapter I show how a new category of "targeted therapies," entailing a new market and a new scientific field, comes to be constructed collaboratively across the fields of business, science, and commerce. To do so I switch perspective. So far, the analysis has focused on stories told by companies themselves in press releases, by media reports, by financial analysts, as well as by a commenting scientific community. In this chapter, in order to understand how economic actors are making sense of this field of breast cancer therapeutics, I analyze the evaluations and interpretations of market analysts, important intermediaries between science, biotechs, and potential investors. This shows the construction of a new category, and, once institutionalized, the second important elements of market emergence from stories.

Evaluating and Interpreting a New Market

Market analysts typically interpret developments of an economic field and evaluate companies and their products. Research has shown their role as central intermediaries: they are "promissory organizations" that simplify complexities by framing situations and naming expectations, thus offering classifications and configurations of products or practices (e.g., Martin 2015; Pollock/Williams 2010, 2016). Moreover, analysts play an important role as "conversation makers," "who can develop interesting perspectives and opinions about a firm or an industry" (Giorgi/Weber 2015: 357), and thus provide a framing. I build on such studies and focus in particular on analysts' attempts to make sense of the new field in evaluating and categorizing. I view market analysts as professional evaluators of particular economic fields. They provide detailed analyses and offer assessments of what is happening in the present and project future developments in form of book-length reports, which often center on statistics and graphs. Such reports offer insight into how they, in their interpretation of the current state of the market and in their evaluation of the future, are trying to make sense of a moving market of breast cancer therapeutics—a market full of expectations and stories of the future.

This chapter brings together twenty-three reports from the market research company Business Insights, published between 1999 and 2011.[1] In these reports, entitled, for example, *The Cancer Outlook 2000* or *Cancer Market Outlook to 2009*, analysts, typically experts in a biochemical field, evaluate the current state of the market, provide forecasts, discuss important companies, and highlight market opportunities. These reports, in turn, are used as data sources in other publications, e.g., *Nature Reviews Drug Discovery*. Data used in these reports include interviews with companies' investor relations and R&D departments.

Hewing to a formula, they usually begin by delineating the "patient potential" for the predominant cancer types, i.e., the number of cancer incidences, the number of patients by stage of cancer, and the state of current treatments. They then turn to promising compounds in the development pipeline to delineate expectations about the future and provide forecasts.

In the late 1990s, market analysts see a growing market for what they then call "innovatives." While sales of chemotherapeutics were forecast to stagnate and hormonals to decrease, "innovatives" were forecast to double in sales, albeit at much lower figures (Marshall 2000). By 2010, by which point "innovatives" have come to be categorized as "targeted therapies," they have a global market share of over 60% annually, with the top four products accounting for more than US$ 20 billion in sales (Business Insights 2011: 65). The creation of a new category, "targeted therapies," is especially evident in reports from market analysts, who, together with the science community, try to make sense of what this new market could be about as it is emerging in the years surrounding the turn of the twentieth century.

This new category marks a shift in the pharmaceutical market. While companies such as Bristol-Myers Squibb essentially dominated the chemotherapy market for decades, developments in molecular oncology research gave rise to new global players. Moreover, by 2005, analysts abandoned the idea of a blockbuster in cancer treatment in favor of "targeted therapies," which use increased knowledge about biochemical markers to treat particular groups of patients, bringing about the advent of the so-called personalized medicine. How this category came to be constructed and acquired meaning highlights the second important process in the emergence of a market for innovative breast cancer therapeutics.

In 1999, the year after Herceptin was FDA approved, a chemotherapeutic product and two hormonal therapeutics were the top-selling drugs in the treatment of breast cancer. The *Cancer Outlook 2000* report (Marshall 2000) divides the treatment market for all cancers into four different drug types: cytotoxic, hormonals, innovatives, and adjuncts (i.e., for the treatment of side effects from cancer therapy). Adjuncts, as the calculation of sales indicate, are the most profitable drug type in 1999. Figure 4.1 shows cancer treatment sales for 1995–1999.

The report summarizes its expectation in an image of an arrow, which starts out in the 1960s in the color of a cold blue and points upward toward the twenty-first century and a hot red future entitled "cure?" This arrow indicating

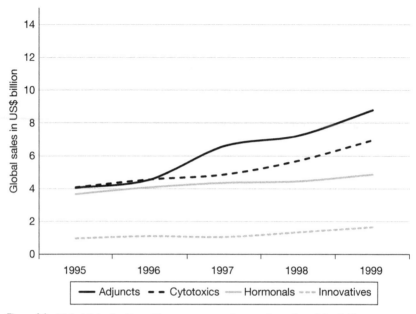

Figure 4.1 Global Sales by Drug Type 1995–1999. Source: Data from Marshall 2000, p. 45.

developments in cancer therapeutics fundamentally captures the inability to foresee how such a future in the "21st century" could look like. It speaks to the uncertainty at the end of the 1990s on what to make of "innovatives" and how to categorize them. The report describes "innovatives" as an "immature market," which consists of "immunotherapies," "which includes both immunomodulators and monoclonal antibodies" (51), and according to the report includes EPO products (see Chapter 2). This groups together research strategies that are typically held apart in the scientific oncological literature of the time, bringing them together here for market analytic purposes. Newer monoclonal antibodies for other indications, the report predicts, "will drive the establishment of the innovative therapies as the fourth major cancer subclass of the future" (51).

In 1999, Bristol-Myers Squibb (BMS) (chemotherapy, hormone therapy), AstraZeneca (hormone therapy), and Novartis (hormone therapy) are the leading pharmaceutical companies for breast cancer therapeutics. The report views Roche as having an expanding oncology portfolio: Herceptin had just started its sales in the United States (1998: US$ 30.8 million; 1999: US$ 188 million).

The approval of Herceptin is evaluated as an important step for the acceptance of monoclonal antibodies. The report also notes (71),

> Herceptin's specificity means that is does not have the toxicity associated with traditional chemotherapy, although its commercial potential (whilst still largely untapped) is reduced due to the fact that approximately only 25% of breast cancer patients express the HER-2 protein.

This evaluation of Herceptin's "reduced commercial potential" suggests the expectation that other future therapeutics would work for greater percentages of patients. It also indicates doubt that an "innovative" drug can become a blockbuster, defined to yield sales of over US$ 1 billion and to be applicable for millions of patients (see also Business Insights 2001).

The 2000 report forecasts sales of chemotherapeutics to stagnate and hormonals to decrease. "Innovatives," however, are seen as having "potential" in terms of revenue and therapeutic benefit, and are expected to triple in sales until 2007 (Marshall 2000: 82). In the chapter entitled "New Market Opportunities," the report notes that current cancer research is changing direction from improving existing therapies to searching for novel compounds. Under "innovative therapies" the report lists anti-angiogenesis agents, immunotherapies, and gene therapies (106). "Immunotherapies are expected to be the most effective," and indeed, the report has great expectations that Theratope "will make a significant impact on cancer therapy in the next 5 to 10 years" (112).[2] In terms of angiogenesis inhibitors, the report primarily discusses MMP inhibitors, which by that time had failed phase III of clinical trials (see Chapter 3).

The 2002 report, *The Cancer Outlook to 2007* (Birch 2002), notes that while "market growth is slowing," "growth in the cancer market is being driven by the innovative therapy class," in which sales grew by 32.5% from 2000 to 2001 (62).[3] In comparison to the 2000 report, the 2002 report adjusts the expected sales figures for "innovatives" upward. The 2002 report also notes that "the current gold-standard for the treatment of breast cancer is tamoxifen," the hormonal treatment for estrogen-positive patients (67) (AstraZeneca). Bristol-Myers Squibb is the leading company in terms of sales (US$ 3.7 billion for all types of cancer), while Roche has the fastest-growing oncology portfolio. Monoclonal antibodies (Rituxan and Herceptin) are showing an "impressive performance"; Genentech's and Roche's sales combined for Herceptin amount to US$ 484 million in 2001 (an increase of 70% from 2000), whereas other "innovatives," such as interferons, are stagnating. The report also indicates that Herceptin is approved for only a "comparatively small subset of the breast cancer population," i.e., 25% of women have HER2 overexpressing tumors. By 2007, the report forecasts, the "innovative cancer therapy class" (175) will

> have overtaken both the cytotoxic and hormonal drug classes in terms of sales. Much of the expansion will occur in currently marketed innovative classes, specifically, the monoclonal antibodies and therapeutic vaccines.

Figure 4.2 visualizes the analyst's sales expectations as of 2002.

As "innovative therapies," the report defines all those "which lie outside of the traditional cytotoxic and hormonal classes" (191). "Innovatives" are thus labeled with great expectations for sales growth, yet there is no clear sense of what this category consists of, as it is a diverse group of "all others." The 2002 report also notes that these products are difficult to evaluate because so much is unknown about their clinical efficacy and because no gold standard exists against which to evaluate them. Indeed, they are a moving target to label.

Another 2002 report on *New Cancer Therapeutics* (Dooley 2002) emphasizes that the "market for cancer is enormous" (18) and that

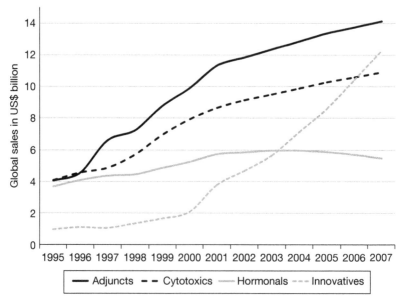

Figure 4.2 Global Sales by Drug Types 1995–2001 and Expected Sales until 2007 (thin lines). Source: Data from Birch 2002.

the science of genomics is set to revolutionize various aspects of oncology practice, including how anticancer drugs are discovered and developed, how cancers are detected and classified, and how patients are treated and monitored.

Unlike prior reports, this report now specifically identifies the mechanism of innovative cancer drugs (36):

> Unlike most currently available cancer therapies, which can have severe toxic effects on normal tissue when used at the dosages necessary to kill tumor cells, the innovative new drugs target particular aspects of the disease, which tends to restrict tissue destruction to the tumor itself. For instance, mAbs work because of their unique specificity for tumor-associated antigens, based on the differences in cell surface protein expression between cancerous and normal tissues. Because many different types of tumors can express an antigen that normal tissues do not, one type of mAb can have potential for many different types of cancer.

The report also identifies various *targets* as "opportunities for molecular-targeting therapeutics for cancer" (33). It also discusses angiogenesis as "a promising target for cancer therapy" and in particular vascular endothelial growth factor (VEGF), which has been identified as a crucial regulator of both normal and pathologic angiogenesis. In some breast cancers an overexpression of VEGF has been identified and anti-VEGF agents are in development, e.g., mAbs that would target VEGF. The report finds *bevacizumab* (called Avastin, developed by Genentech/Roche) to appear "set to be the 'first-to-market' anti-angiogenesis inhibitor for breast cancer" (152).

However, the 2004 report on *The Cancer Outlook to 2009* (Seget 2004) shows enduring confusion about how to categorize these new treatments. They are "novel treatments, which a more targeted in nature" (13), and promise to yield solutions to unmet needs. Herceptin offers "a more specific, and hence less toxic, treatment" (139). And (27):

> Novel therapies are starting to make an impact on the breast cancer market, but none have yet gained acceptance as gold-standard treatments, although Roche/Genentech's monoclonal antibody Herceptin is rapidly moving toward gold-standard therapy for the subset of patients with metastatic disease whose tumors over-express HER2.

While the prior report of the same market research company identified a mechanism of innovative therapeutics to be "targeted," this 2004 report of the field uses "target" frequently without having a concept of a biochemical mechanism in mind. Products are targets for market participants, as in "attractive target for generic manufacturers" (119), yet also, cytotoxic therapy ought to target only malignant cells (87). The report is less enthusiastic about the potential of innovatives as previous ones: they will be as successful in terms of sales as chemotherapeutics, but no new classes of innovatives will drive the market in addition to monoclonal antibodies and therapeutic vaccines.

In sharp contrast, another 2004 report uses the term "targeted cancer therapies" to categorically and collectively describe recent research developments. It even indicates a new business model. Such targeted medicines will "lead to the development of personalized health maintenance programs" as well as to (Mertens 2004: 12)

> blockbuster opportunities in other disease markets where there are fewer therapies and a high degree of unmet need enabling highly innovative products to compete strongly.

Future treatments, this report expects, will depend less on cancer type and more on molecular features (48):

> In the future, the treatment for each patient's cancer will be individualized, based on the unique repertoire of molecular targets expressed by their particular tumor.

The report is also sure to point out that (48)

> [o]ncology is also attractive because this therapeutic avenue has a price component that is boosted as the disease acquires chronic features, a factor of great influence in any company's R&D investment decision.

When patients are treated with targeted cancer therapies, which promise to reduce side effects and increase survival (57), this in turn means that patients can continue to take the targeted treatment—thus ensuring continuous profits for pharmaceutical companies.

The report distinguishes innovatives "specifically targeting cancer cells" (57) by mode of action: there are angiogenesis inhibitors, immunostimulators, and tumor-targeted agents. The report views anti-angiogenesis approaches to be particularly promising cancer treatments. In particular, the anti-VEGF treatment *bevacizumab* (Avastin) looks especially encouraging for breast cancer treatments. Herceptin is "rapidly moving toward gold-standard therapy" for

25–30% of breast cancer patients in which HER2 is overexpressed (83). But mutation of the p53 tumor suppressor gene, the report points out, can be found in "up to 50% of all primary breast carcinomas" (84); this then presents a market opportunity.

By 2004, and after three cancer therapeutics that target and bind to particular enzymes have received regulatory approval,[4] molecular oncological R&D projects in the clinical pipeline account for 29% of all R&D activity and are the largest investments of pharmaceutical companies (37). Now, "targeted therapies" is a label for a class of treatments in comparison to others ("chemotherapies, haematopoietic factors of growth, hormonal therapies"). The expected sales of targeted therapies for the global oncology market almost doubles from US$ 9.2 billion in 2004 to US$ 17.5 billion in 2008 (Mertens 2005: 66). No other drug class receives such high sales expectations.

As the report states, Roche has become "a key leader in targeted cancer therapies" (Mertens 2004: 123) by 2004, with "a strong cancer pipeline, partly attributable to its collaboration with Genentech" (16) as well as several strategic alliances with companies developing targeted therapeutics. Genentech is focusing on targeted therapies only. AstraZeneca also has a strong oncology portfolio of targeted treatments, of which one looks promising to treat breast cancer (the EGFR inhibitor *gefitinib*, called Iressa). The formerly leading oncology company Bristol-Myers Squibb, however, is stalling in growth.

This shift to targeted therapies marks a watershed moment for the pharmaceutical industry. The reports speak of a "new business model," in that big pharma companies will need to form a greater number of strategic and licensing alliances to "spread the risk of R&D pipeline failures" (Mertens 2004: 157). Substantively, targeted cancer therapeutics also suggest that treatments for smaller, more specific populations can be developed. These changes in biochemical knowledge occur at a time when the business model of a "blockbuster drug," i.e., a drug with more than US$ 1 billion in sales and meant for millions of patients, has come to be unsustainable: R&D productivity had been falling and no blockbuster candidate appeared in sight. In turn, analysts see the move toward developing specialist drugs and personalized medicine (Mertens 2005: 15) and particularly cancer treatments as "the golden future of healthcare" (16).

Opposing a "new millennium model" of targeted medicine to a "broadband model," the report *Beyond the Blockbuster Drug* (Mertens 2005) also suggests a growing medical need and thus growing market segment for "profiling" on the basis of genetic tests. While some mutations in breast cancer susceptible genes (BRCA1 and BRCA2) have been identified in the early 1990s to lead to an increased probability of developing familial breast cancer, these mutations only hold for 5–10% of all women developing breast cancer. "Profiling" also refers to the field of biomarkers and target diagnostics, in which particular receptibility, e.g., ER+ or HER2+, can be determined (e.g., Baker 2005; Carney 2005; Drews/ Ryser 1997; Silber 2001).[5]

The 2005 report's prediction is that the era of medicines for "every" body is over; instead we are moving toward "medicines for 'my' body" (127). And (emphasis in original, 127):

> As a result, profit margins can only grow. *It makes the future extremely exciting.*

Moving forward a couple of years, the *Cancer Market Outlook to 2016* report (Business Insights 2011) distinguishes between six key drug classes for all cancer treatments, namely *targeted therapies* (i.e., monoclonal antibodies, angiogenesis inhibitors, tyrosine kinase inhibitors) in addition to chemotherapeutics, immunotherapy, adjuncts, chemo-prevention, and blood cell mobilization (89). By 2010, monoclonal antibodies account for the most sales, "achieving a 30.7% market share among the top 10 brands in the global cancer market" (64). The top four products (Avastin, Herceptin, Rituxan/MabThera, and Gleevec) are targeted therapies, mAbs and small molecules, and account for more than US$ 20 billion in sales (65).

In the United States, another study shows, sales of targeted therapies have outperformed sales of chemotherapies (see Figure 4.3). Industry analysts now speak of a "market for targeted cancer therapies," which has been growing amid an otherwise slowing pharmaceutical market (Aggarwal 2010). Growth also results from an expansion of indications—originally approved to treat one type of cancer, products also gain approval for other types of cancer with the same target.

By the end of 2010, three targeted therapeutics to treat breast cancer have been approved. In addition to *trastuzumab* (Herceptin), *lapatinib* (Tykerb/ Tyverb) and *bevacizumab* (Avastin) have received FDA and European Medicines Agency (EMA) approval and are part of that market.

Lapatinib (Tykerb/Tyverb) is a small molecule that inhibits activity of HER2/neu and of the epidermal growth factor receptor type 2 (EGFR2). Developed by GlaxoSmithKline (London, UK) as a treatment for solid tumors such as breast and lung cancer, with first clinical trials in 2004, it received FDA

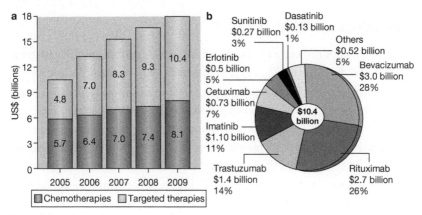

Figure 4.3 Representation of a "Market for Targeted Cancer Therapies" based on US Sales. Source: Aggarwal 2010, p. 427.

approval in March 2007 and EMA approval in June 2008, for use in patients with advanced metastatic breast cancer in conjunction with a chemotherapeutic agent. *Lapatinib* responds particularly to women who are tested positively for estrogen receptors (ER+), epidermal growth receptors (EGFR+), and human EGFR type 2 (HER2+)—and are, so called, "triple positive" (Moy et al. 2007).

Bevacizumab (Avastin) is a humanized monoclonal antibody directed against vascular endothelial growth factor A (VEGF-A), a protein that helps tumors form new blood vessels (angiogenesis inhibitor). It was developed by Genentech/Roche and received first FDA approval in 2004 for the treatment of colon cancer (Ferrara et al. 2004; Muhsin et al. 2004). Provisionally approved by the FDA in 2008 for metastatic breast cancer, this approval was revoked in November 2011. In well-publicized meetings, the FDA found that although *bevacizumab* seemed to slow cancer growth for a short time in some women, it did not help them live longer and rather posed health risks (Couzin-Frankel/Ogale 2011). However, Avastin has been EMA approved since January 2005 and is still available for breast cancer treatment in the European Union and, through other indications, as off-label breast cancer treatment in the United States.

As the *Cancer Market Outlook to 2016* summarizes (Business Insights 2011: 85):

> Targeted therapies are revolutionizing the paradigm of cancer treatment and are likely to be used in most cancer patients in the next 10 years.

A New Category: Targeted Therapies

Just as companies and their financial analysts before, market research analysts tell stories of the future and communicate their expectations. On top of focusing on the worth of individual products, molecules, and companies, such reports also show how categories come to be constructed from stories of the future. In trying to make sense of what is going on in the emerging market and to provide analysis and forecast, market research analysts are involved in labeling products and categorizing them into larger entities to make them cognitively accessible to interested investors. While doing so, they themselves are searching to make sense of what is going on.

Research on categorization processes has pointed to different phases in the emergence of new categories (e.g., Bingham/Kahl 2013; Grodal et al. 2015). At the core rests the idea of a seemingly paradoxical move of "making the unfamiliar familiar but conceptually distinct" (Bingham/Kahl 2013). In case of new breast cancer treatments, the conceptual distinction is aided by fundamental biochemical distinctions. *Targeted therapies* are technologically and categorically distinct from chemotherapies; categories and the underlying technology had co-evolved. However, precisely *how* the biochemical processes and, even further, new molecules and compounds function remain contested and indeed uncertain. The following thus extends the ideas of co-evolution of categories and technologies, introduced in the introduction, to an example for unknown unknowns in market, products, and processes.

The establishment of the new category *targeted therapies* and the associated focus on *targeted therapeutics* underwent three phases from changing labels to institutionalization: first there was a shallow label, followed by a new label for a category that bridged novelty with familiarity, which raised confusion but also turned out to serve as a "trading zone" to which different actors could attach their interpretations and meaning. The flexibility of the new category allowed for its institutionalization.

At first, there was a label, "innovatives," that enabled analysts to communicate new therapeutic approaches. Yet as a residual category capturing a diversity of approaches, its meaning was shallow. As some of the treatments grouped in this category came to be commercially successful, market analysts remained confused and challenged how to categorize such new breast cancer treatments. The category label "innovatives" signaled novelty and high expectations, while market analysts remained uncertain about the ramifications of such products.

Their measures of evaluation were proven standards in treatment from the prior forty years.

In 2002, the label "innovatives" gets boldly replaced with "targeted therapies," a term that captures both novelty and familiarity. As a compounded term and label, it is (Grodal et al. 2015: 429)

> (1) distinctive enough to convey the novelty of the underlying product and attract the attention of stakeholders and (2) familiar enough to be easily comprehensible.

It connects to preexisting categories of "therapies" while highlighting a clear difference of being "targeted" as opposed to unspecific, toxic chemotherapies. Yet here, too, this new category leads to confusion and rejection from the analysts, who are still uncertain what to make of the new treatments. Indeed, all sorts of targets exist in cancer treatment (Seget 2004), a view that turns the category *targeted therapies* into a shallow category, just as "innovatives" was before. While some analysts are involved in sensegiving activities as to what this new category entails, such as the move to personalized medicine (Mertens 2005), others are confused during that phase.

This confusion of market research analysts about how to interpret and evaluate targeted therapies as a new treatment category is mirrored in public discussion of science and biotech analysts, who are involved in "iterative sensemaking, debating and discussing the meaning" of the proposed category (Grodal et al. 2015: 431). Scientific discussions see both potential and challenges attached to the new category. With an increased knowledge about mutant genes (Weinberg 1996) and biological mechanisms of cancer (Hanahan/Weinberg 2000), cancer research since the late 1980s has discussed new molecular targets that could restore normalcy or kill tumorous cells without harming healthy ones. Both clinically successful products and insights into cancer mechanisms shaped the scientific outlook toward the role of targets and targeted therapies as the therapies of the future (Oliff et al. 1996). By 2002, several monoclonal antibodies had been approved for a variety of medical conditions, recognizing particular molecular targets (e.g., Herceptin recognizing HER2), so that a *Nature* article claimed "magic bullets hit the target" (Gura 2002). Selective targets allowed for developing "bullets instead of sledgehammers" for diseases (Atkins/Gershell 2002).

Yet while principal mechanisms had been identified and indeed proven for particular targets, targeted *therapies* were not as easy to develop and to turn into medical products as was first expected. Significant limitations existed, which weakened scientists' initial excitement about the new treatment category *targeted therapies*. Indeed, was this really a category or just a couple of chance products? While molecular biology was able to identify many targets, it proved difficult to turn them into cancer treatments because of existing technical and biochemical limitations (Gibbs 2000). At that time several hundred molecular oncology compounds were in development, of which only few achieved regulatory approval. Because their biological setup makes them "intrinsically complicated" to produce and to use, monoclonal antibodies presented themselves as the "king in the kingdom of uncertainty" (Anonymous 2005). Also, which target to select for drug discovery became a growing concern (Knowles/Gromo 2003).

In the early 2000s, several different definitions of *targeted therapies* are noticeable in the field: the FDA regards targeted therapy as a drug with a previously approved diagnostic test (Ross et al. 2004), as done with Herceptin. Many scientists and oncologists regard targeted therapy to be a "drug with a focused mechanism that specifically acts on a well-defined target or biologic pathway" (599). Other scientists consider anticancer antibodies that "seek out and kill malignant cells bearing the target antigen, as an additional type of targeted therapy" (599).

Discussions in the field of biotechnology echoed the wavering scientific discussions about the role of targeted therapies during the early 2000s. Here, too, expectations were disappointed. Some saw molecular products as unable to attain financial blockbuster status, a drug with expected sales of US$ 1 billion, in the near future (Business Insights 2001). Others forecast that targeted therapies were to reshape the blockbuster *model* for pharmaceutical companies, which meant relying on a few highly successful drugs for large groups of patients to sustain growth (Frantz 2005). With targeted therapeutics, the discovery of drugs had become more complex, more expensive, and also more specific: while R&D costs were skyrocketing, "99.9% of the compounds wash out of the development pipeline" (Service 2004: 1797), and validation for particular molecular profiles takes more time than expected. Clinical trials are also more complicated and in turn regulatory changes are necessary for trial setups (Booth et al. 2003).

Despite these uncertainties and doubts, the biochemical nature of targeted therapies coupled with the dearth of innovations and stagnating productivity (Booth/Zemmel 2004) led many pharmaceutical companies to move "from a blockbuster model to a targeted model." Instead of developing blockbuster drugs, which "are so last century" (Berenson 2005), pharmaceutical companies shifted their attention to focus on specific research areas and personalized medicine. Approved targeted cancer therapeutics Herceptin and Gleevec were seen as harbingers of a larger shift to personalized medicine (Ginsburg/McCarthy 2001; Hedgecoe 2004). Such strategy of "the right drug for the right person" also required advances in diagnostics and biomarkers (Allison 2008).

The category *targeted therapies* not only captures the biochemical process of "targeting," while remaining comprehensible with a reference to therapies. As a category, it also appears flexible enough to connect a variety of interpretations and expectations of worth. For some, this category speaks to the need for new diagnostic tests for particular drug targets, as the FDA first described it (Ross et al. 2004); for others, it captures the end of the pharmaceutical blockbuster model and instead, a turn to personalized medicine, which will have to deal with the uncertainties involved in target selection (Knowles/Gromo 2003). Yet again, others see a great medical potential and the promise to develop nontoxic, noninvasive cancer treatment that fits to a better genomic understanding. For still others, the category *targeted therapies* epitomizes the success of biotechnology as it combines science, finance, and industry.

The category *targeted therapies* therefore serves as a *trading zone* (Galison 1997): it means substantively different aspects to different actors, yet they all can agree that it is about a common concept, "targeted therapies." According to Galison's notion of a trading zone, "the trading partners can hammer out a *local* coordination, despite vast *global* differences" (783). They do so by developing a "pidgin" for these negotiations, which allows them to agree about the concept temporarily and create an understanding of what it means—they collaboratively construct the category *targeted therapies*. For the members of the emerging market, the category *targeted therapies* serves as a quality and a focal point of interest to which they can attach their meaning.

In relating stories of how this new treatment category was to be interpreted and evaluated, a diversity of actors from the fields of market research, science, and the biotech industry make sense of what is going on in the present. Just as in the mechanism previously identified in companies' stories, the different actors tie a present state of knowledge to expectations in the future. They do so collaboratively, as they make sense in relation to how others are making sense of these new molecules and treatment approaches. As a trading zone of discussions and contestations, the new category *targeted therapies* gets "domesticated" or institutionalized to move from the unfamiliar to the familiar. Its flexibility was essential for its institutionalization as a category in the second half of the 2000s.[6]

A third phase of the categorization process thus is its institutionalization in the second half of the 2000s. By then, stories of the future no longer only refer to particular molecules, compounds, biochemical mechanisms, or companies, but also to this new treatment category.

Once the category *targeted therapies* is institutionalized, however, it does not remain uncontested. After Herceptin's FDA approval in 1998, nine years passed before another molecular treatment for breast cancer received approval. Many expectations had turned into disappointments by then. As a category, *targeted therapies* brought into sharp focus scientific problems of cancer models as such, which indeed continue to affect progress in finding treatments: "differences among the therapeutic agents in patient progression and survival rates are mysterious" (Kamb et al. 2007: 117), classification of tumors varies from pathologist to molecular researchers to clinical oncologist, and some of these classifications may even be "a myth" (118). Yet as a category for a new market of targeted therapies, it successfully stuck.

5

Patterns in Meaning-Making
Categories over Time

THE FIELD OF INNOVATIVE BREAST cancer gave rise, as we saw in the previous chapter on a qualitative inquiry into categorization processes, to the idea of *targeted therapies*. As a category, the concept of *targeted therapies* was precise enough to fit a new type of treatment, yet also flexible enough for many different actors so they could bring forward their associations and attach their meaning. Over time, however, the associations and meaning attached to this compounded term concretized. This process of finding first a label and then collaboratively institutionalizing a category represents the second important element for the emergence of a market from stories.

The ensuing empirical analyses delve into the collaborative construction of categories even further. After close readings of the stories of companies, financial and industry analysts, as well as journalists during the emergence of the market and the institutionalization of the new treatment category *targeted therapies*, this chapter now zooms out and examines developments of those entire fields from a macroscopic perspective over time. To do so, it uses topic modeling and semantic network analyses, methods that allow for identifying discursive trajectories as well as drifts and shifts of categories over time. These methods also allow us to study stories of multiple actor-networks. The analyses use four different textual corpora—abstracts of oncological expert discussions, industry newsletters, wire reports, and newspaper reports—to study processes of meaning-making and to identify categories over a twenty-two-year-period. The goal is to obtain a better understanding of the emergence of a market, including the construction and then institutionalization of *targeted therapies* as a category across different, though interconnected fields. What are the trajectories of different biochemical mechanisms, compounds, and molecules over time? Also, what role do objects such as compounds and molecules play in the stories overall? Topic modeling and semantic network analyses provide answers to such questions.

The central analytical idea this inquiry now rests upon is that words acquire meaning in relation with other words. Such associations are typically more than a pair of compounded terms and rather a cluster of co-occurring terms that together make meaning. Based on co-occurring terms, categories arise.

What is of interest then is not a mere counting of mentions, such as in frequency counts of, say, "target" over a period of time, but rather the co-occurrence of "target" with other terms. To be sure, terms have multiple meanings, e.g., "target" refers to a new treatment category, as previously discussed, but might also refer to the target of a stock price. Moreover, key terms co-occurring with "target" may shift over time, and this is especially evident in a scientific field such as cancer research, in which new knowledge surpasses old. And although the terms may shift given new knowledge, the substantive category of discussion may remain the same. "Effector-to-target cells" of 1989 and "target genes" of 2010 may belong to the same category of co-occurrences over time. To detect such categories and their shifts, *a priori* codes that rest on either a basic understanding for what is going on in the text (e.g., Krippendorff 2012; Kuckartz 2004, 2007) or on dictionary-based models that compare terms in the corpus with those in a predefined list (e.g., Grimmer/Stewart 2013) are thus not useful. In order to detect categories of discussions and to gain insights into their trajectories across time, I apply two different types of analyses: topic modeling on all four corpora as well as semantic network analysis on the corpus of scientific discussions. These methodological approaches are able to capture term co-occurrences and to track these co-occurrences over time. While topic modeling is a useful method to discover categories from a corpus of texts over large periods of time, the semantic network analysis employed in the second half of the chapter can show drifts and shifts in the structure of terms making up the categories over time using network analytic algorithms. It will be used to track the specifics of *targeted therapies*, which can most fruitfully be analyzed in the scientific discussions. Overall, these analyses show how the category *targeted therapies* was shaped and became institutionalized over time and collaboratively in the fields of science, the biotech industry, and journalism. These stories of meaning-making are the context in which the nascent market for new breast cancer therapeutics emerges.

Topic Model Analyses: Trajectories of Meaning over Time

Developed in the fields of computer science, machine learning, and natural language processing (Blei 2012; Blei/Lafferty 2009; Blei et al. 2003), topic modeling has received increased attention in recent years in the humanities (e.g., Goldstone/Underwood 2014; Graham et al. 2016; Jockers 2013; Meeks/Weingart 2013; Moretti 2013) and the social sciences (e.g., Bail 2016a; DiMaggio et al. 2013; Evans/Aceves 2016; Grimmer 2010; Grimmer/King 2011; Grimmer/Stewart 2013; Kaplan/Vakili 2014; Karell/Freedman 2019, 2020; Light/adams 2016; McFarland et al. 2013; Mohr/Bogdanov 2013; Mohr et al. 2013; Mützel 2015; Nelson 2021; Ramage et al. 2009). Topic models provide an automated procedure for sorting the contents of a text corpus into a set of meaningful "topics," in which terms jointly appear unusually often, forming a category.[1] Topic modeling algorithms involve minimal human intervention: researchers need to specify a number of topics for the algorithm to find within the given corpus; the algorithm then yields that number of topics and the words being used in a topic as well as the distribution of those topics across the text corpus. Here I work with Latent Dirichlet Allocation (LDA), a common and simple topic model.

Substantively, the idea is that a collection of texts exhibits several topics, in which particular words sort together because they co-occur unusually often. Topic modeling assumes that the order of words in the document does not matter. While automatic coding for frequencies finds the spread of terms across an entire textual corpus, e.g., the term "target" used most frequently, frequency counts cannot provide automatically for relational meaning of terms co-occurring in similar constellations in documents. Topic modeling works with this idea of words' relational meaning within a document, an idea similar to those developed in network analytic studies using texts for delineating deeper thematic structures (e.g., Bearman/Stovel 2000; Moody/Light 2006; Smith 2007). Moreover, topic modeling is able to capture multiple meanings of a word: a word may belong to different topics, embedded in different meaning structures in each. Thus topic modeling looks for "deep structure" and patterns of meaning in a corpus of documents as it clusters together semantically related words.

More technically, topic models are fully automated clustering techniques, allowing for mixed membership of terms, analyzed at the document level. LDA assumes that there are a number of topics in a collection of texts, which can be discovered. A topic is a group of words that co-occur with each other unusually often across all documents. Each document exhibits topics with different proportions. A newspaper article may be about a new cancer treatment, profitability of a company, management strategy, and patients' health during treatment, all in the same article, though with different proportions. Computationally, topic modeling finds the hidden structure that likely generated the observed collection of documents with their words, in a reverse generative process. To do so, it uses Bayesian statistical techniques and infers, given all documents, what the topics are that each document draws on. LDA optimizes for two goals to find "good topics": "first, for each document, allocate its observed words to few topics; second, for each topic, assign high probability to few words from the vocabulary" (DiMaggio et al. 2013: 578).

The inferred hidden structure resembles the thematic structure of the collection. The LDA algorithm yields distributions of topics over the corpus of documents as well as lists of top terms making up topics. Thus it generates topics from the documents rather than imposing categories or keywords *a priori* on the texts and provides for a way of tracking the evolution of topics over time over large bodies of text. It turns out that the resulting topics are interpretable, and indeed may be viewed as "frames (semantic contexts that prime particular associations or interpretations of a phenomenon in a reader) and employed accordingly" (DiMaggio et al. 2013: 578). Topic modeling thus presents an "inductive relational approach to the study of culture" (576) based on formal analysis. The semantically cohesive patterns detected and grouped as topics offer descriptive explanations of shifts in content in a given field. To be sure, topic modeling only moves substantive human interpretation to a later position in the analytical process—it does not replace it. Analysts need knowledge of the respective field to interpret the resulting topics.

One of the methodological issues raised in discussions on topic modeling is how to validate the results. In a systematic discussion of using topic models, Grimmer and Stewart (2013) emphasize the need for topics to be labeled. They suggest two methods for validating topic labels. One evaluation is based on semantic validity: do the topics identify coherent groups that are "internally homogeneous yet distinctive from other topics" (Grimmer/Stewart 2013: 287)? Semantic validation can be obtained by having other readers check the topic labels—in my case, I asked cancer researchers to look over the topic labels. One other evaluation is based on predictive validity. Quinn et al. (2010) argue that if topics are valid, the external events should explain sudden increases in attention to a topic. This too my analysis will show.

However, topic model analysis, especially in this simple form, has its limitations. The bag-of-words assumption disregards the order of words within a text. Grammar, rhetoric, and sentiment cannot be taken into account. Also, the order of documents does not matter. The topic model algorithm produces lists of ranked terms according to the "good topic" optimization, while ignoring the structure of co-occurrence of words and their dynamic over time. After the algorithm has sorted the corpus, interpretive work is necessary to read the entire amount of topics and to think about the phenomena under scrutiny. There is no correct number of topics in a corpus; this stands in sharp contrast to other statistical approaches. As DiMaggio et al. (2013) put it: "The point is not to estimate population parameters correctly, but to identify the lens through which one can see the data most clearly" (582).

Another limitation of LDA topic model algorithms is that they are not able to model topics arising from another topic or to merge with another (but see Cui et al. 2011).

In the ensuing analysis, topic modeling is used as a heuristic tool for the tracing of content trajectories in the field of breast cancer therapy using four different textual corpora. In bringing the different fields of science, industry analysis, company reports, and journalism together in one analytical framework, particularities as well as similarities and overlaps become evident from a macroscopic view.

Topics in Scientific Discussions, 1989–2010

A first inquiry is into scientific discussions. The data set consists of all abstracts on breast cancer therapy in the subject area of oncology from the Web of Science database.[2] This represents the state of the art of published research findings on breast cancer therapeutics over twenty-two years. Whereas the other three large-scale corpora emphasize expectations and future possibilities, documents in this corpus present findings.

The search yielded a total of n=31,070 documents, of which n=30,139 abstracts were usable for further analysis; especially in the early 1990s, several entries came without abstracts. Figure 5.1 shows the distribution of documents and abstracts over time.

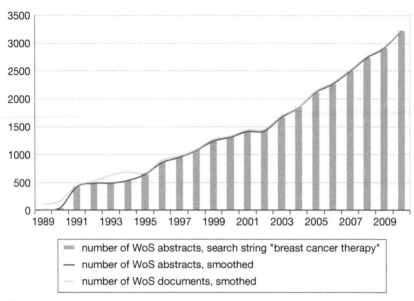

Figure 5.1 Distribution of Web of Science Documents and Abstracts, 1989–2010

There were 221 journals contributing to these discussions. Most contributions came from the journals *Cancer Research* (n=1,858) and *Clinical Cancer Research* (n=1,651). Several journals had only one contribution.

In preparatory steps, I cleaned the documents and removed single numbers, sentence markers, stop words, the most frequent words, and those most rarely used. I ran the LDA topic modeling algorithm with Gibbs sampling[3] on 30,139 abstracts to show the development of topics, i.e., terms that unusually often cluster together. I ran different models to establish the number of topics. I ended up with 200 topics that are distributed over all abstracts between 1990 and 2010 (twenty-one years) as the most robust and interpretable model.[4]

The analysis of the abstracts shows the development of topics within the entire corpus of scientific discussion over time. The model shows different topics across all years, such as general discussions about health care; quality of life; regimen dosage; particular types of cancer (e.g., neck, lung, ovarian); metastases; research centers; and clinical trial results.

One of the core tensions and indeed the initial backdrop for the emergence of a new market for cancer therapeutics is the tension between toxic and targeted treatments. The results of the topic model analysis (Figure 5.2) show the distribution of topics across all abstracts of the topics I have labeled "role of targets" and "toxicity, chemo regimen." The topic on toxicity in chemotherapeutic treatments was the most prevalent topic of all topics in the entire corpus and the most prevalent in 1990, when it was part of 6% of all abstracts. By 2010 it had clearly decreased, contrary to the topic of "role of targets," which was steadily rising to become the biggest and most prevalent topic in 2010. The shift in topics occurs in 2002. This matches the prior, qualitatively gained insights, that beginning with 2002 *targeted therapies* was introduced as a label for a new treatment category.

Similarly, terms such as "new, novel, developments, approaches, treatment, target" indicate the rise of new approaches in the field of breast cancer therapeutics. This topic on "new/novel approaches" was the second most often found topic in 2010. When added to the previous topics, it similarly illustrates a rise vis-à-vis the topic on "toxicity, chemo regimen" (Figure 5.3). This topic model analysis of Web of Science abstracts thus makes evident that a topic of "targeted therapies" exists in the scientific discussions and indeed, that the topic increases in occurrence over time. This only supports the finding that a new treatment category of *targeted therapies* stuck and helped to make sense of what this new market was about.

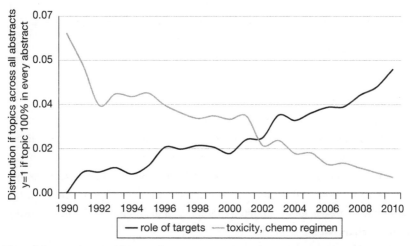

Figure 5.2 Distribution of Topics "Role of Targets" and "Toxicity, Chemo Regimen" across Web of Science Abstracts, 1990–2010.

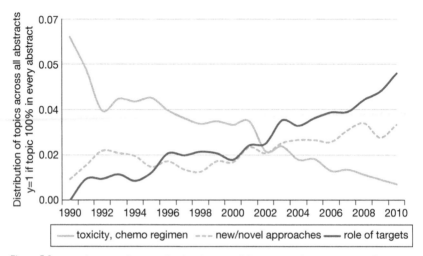

Figure 5.3 Distribution of Topics "Role of Targets," "Toxicity, Chemo Regimen," and "New/Novel Approaches" across Web of Science Abstracts, 1990–2010

Table 5.1. 25 Top Terms of Three Top Topics, Web of Science Abstracts, 1990–2010

toxicity, chemo regimen: toxicity, grade, mg/m, vinorelbine, weeks, cycles, toxicities, days, advanced, range, every, complete, partial, regimen, responses, day, progression, administered, occurred, evaluable, duration, efficacy, phase, epirubicin, given

role of targets: role, its, show, signaling, proliferation, progression, important, can, activation, target, through, here, function, mechanisms, mechanism, novel, development, pathway, cancers, cellular, pathways, including, potential, regulation, molecular

new/novel approaches: new, will, development, therapies, molecular, such, targeted, potential, understanding, agents, review, strategies, research, recent, can, important, novel, most, approaches, many, several, mechanisms, future, current, specific

Table 5.1 illustrates how the topics are constituted. It lists the 25 top terms making up each topic; the name of each topic is listed in bold. Note that while the LDA algorithm delivers a ranking of terms according to probability distributions as a group, the names of the topic are my own.[5] As could be gathered from previous qualitative analyses, "targets" are about mechanisms, pathways; their role is novel and important. The topic "new/novel approaches," on the other hand, speaks about "targeted therapies." It also refers to the future, using terms such as "will, potential."

From the study of the relevant literature and as earlier analyses of stories of biotechs, financial and industry analysts, and market analysts have shown, these novel approaches can be grouped into different research strategies and clustered around biochemical mechanisms, compounds, or molecules. Figure 5.4 shows topics that are pertinent to analyses of earlier chapters while showing their usage over twenty-one years. Figure 5.4 also indicates when innovative, targeted therapeutics for breast cancer received FDA approval.

While the scientific discussion on interleukins diminishes in the late 1990s, discussions on genetic mutations and a particular mechanism to kill cancerous cells from within the cell (apoptosis) increase. Similarly, discussions on HER2, the receptor to which Herceptin, approved in 1998, binds, increase in the early 2000s. At the end of the period studied, the strongest discussion is on mTOR inhibitors, which are viewed as able to treat HER-negative breast cancer, which is otherwise unresponsive to hormonal or other receptor treatments. Until 2011, no treatment for those breast cancer types had been approved. Also visible is a rise of an MMP topic, the mechanism that so grandiosely failed in clinical trials in the United Kingdom, while the underlying biochemical mechanism is undisputed and was discussed again in the late 2000s. The topic of immunotherapy also had a peak in the early 2000s, just before the first preventive vaccine for cervical cancer was approved.

Table 5.2 displays 25 top terms that form 9 selected topics out of the 200-topic solution the Gibbs LDA model yielded. Again, the topic titles in bold are my own. The topics on mechanisms, molecules, and compounds also highlight the role of nonhuman actors in this field.

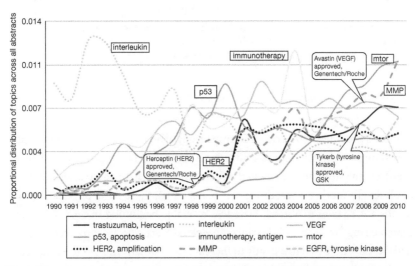

Figure 5.4 Selected Topics on Biomedical Mechanisms, Molecules, and Compounds with Information on Events Taking Place, Web of Science Abstracts, 1990–2010

Table 5.2. 25 Top Terms of Nine Topics on Selected Mechanisms, Targets, and Compounds, Web of Science Abstracts, 1990–2010

trastuzumab, Herceptin: trastuzumab, her2, her2-positive, herceptin, antibody, epidermal, her2-overexpressing, monoclonal, her2/neu, anti-her2, pertuzumab, her3, overexpressing, her, mg/kg, overexpression, alone, plus, humanized, her2-negative, positive, bt474, trastuzumab-based, her4, lapatinib

p53, apoptosis: p53, mutations, mutation, mutant, wild-type, status, tp53, pik3ca, mdm2, mutated, dna, p21, alterations, suppressor, cancers, found, sequencing, k-ras, type, exons, apoptosis, function, wild, presence, somatic

HER2, amplification: her2, amplification, ihc, fish, status, situ, overexpression, her2/neu, hybridization, immunohistochemistry, cases, fluorescence, top2a, positive, testing, samples, concordance, negative, copy, number, specimens, amplified, determined, herceptest, found

interleukin: il-2, cytotoxicity, lymphocytes, immunotherapy, immune, vitro, killer, ifn-gamma, lysis, adcc, dcs, effector, activated, against, peripheral, blood, lak, natural, monocytes, autologous, mononuclear, cytotoxic, til, il-4, interleukin-2

immunotherapy, antigen: antigens, antigen, peptides, peptide, ctl, immunotherapy, class, responses, specific, epitopes, immune, against, t-cell, mhc, expressed, recognized, cd8, ctls, cytotoxic, target, hla, autologous, vaccination, lymphocytes, induced

MMP: invasion, migration, metastasis, matrix, adhesion, invasive, e-cadherin, mmp-2, mmp-9, motility, integrin, invasiveness, progression, extracellular, mmp, mmps, emt, role, timp-1, potential, vitro, phenotype, metalloproteinase, reduced, matrigel

VEGF: vegf, endothelial, angiogenesis, vascular, angiogenic, density, antiangiogenic, microvessel, vessels, mvd, blood, vessel, factors, lymphatic, bfgf, metastasis, endostatin, anti-angiogenic, formation, vasculature, vegf-c, bevacizumab, antibody, cd31, vegf-a

mTOR: mtor, rapamycin, pathway, akt, target, pten, pi3k, mammalian, signaling, activation, inhibitor, kinase, inhibitors, inhibition, phosphorylation, rad001, everolimus, downstream, phosphatidylinositol, eif4e, 3-kinase, ampk, pakt, activated, pathways

EGFR, tyrosine kinase: egfr, epidermal, gefitinib, lapatinib, tyrosine, kinase, inhibitor, erlotinib, inhibitors, family, zd1839, signaling, inhibition, receptors, therapies, targeting, iressa, members, kinases, erbb, dual, targeted, phosphorylation, inhibited, downstream

Topics in Industry Analysts' Reports, 1992–2010

The second large-scale text corpus contains reports from "dedicated biotechnology news providers" (Morrison/Cornips 2012), who translate and communicate "the future-oriented technical claims for a somewhat heterogeneous audience" (266) of investors and interested others. In contrast to the earlier-analyzed book-length market research reports (Chapter 4), these news providers create a constant buzz of newsfeed. Data was available beginning in 1992, when the search string "biotech* AND cancer*" yielded 8 articles. In 2002 that number had steadily increased to 43, only to increase substantially in 2003 to 817 articles. The topic model analysis is based on n=93,887 reports over nineteen years. Figure 5.5 shows the distribution of newsletter documents across all years.[6]

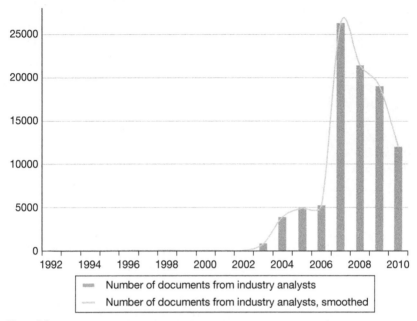

Figure 5.5 Distribution of Newsletter Documents from Industry Analysts, 1992–2010, *Pharma Marketletter* and *Biotech Business Week*

For this corpus a 200-topic solution also yielded robust results. The range of topics was rather wide; it included, e.g., clusters on adverse reactions to treatment, grants received, genetics, dietary intake, health care and prevention, and country-specific topics (for the United States, Korea, Taiwan, and Denmark, among others). As in the previous section, here I select several topics for heuristic purposes.

In the early years of the analyzed period, the topic I have labeled "oncology, market, expect, growth" was the single most dominant topic of all. This topic clusters analysts' discussions on how the industry will grow based on products in the development pipeline (see Table 5.3 for the 25 top terms making up each selected topic). It declines toward the end of the time period analyzed. After many failed expectations, analysts changed how they spoke about the market. Another dominant topic in the early years is on "FDA approval," which is followed by "company, stock prices," a topic that speaks to companies' market activities. A topic on "drug discovery," capturing unusually often co-occurring terms such as "compounds, drugs, oncology, candidates," rises in particular in the early 2000s, when the topic on market expectations is decreasing.

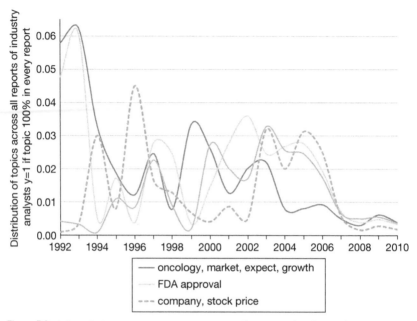

Figure 5.6 Selected Topics on Economic Stories, Industry Analysts' Newsletters, 1992–2010

Evidently, industry analysts shifted from a focus on expectations to a focus on actual compounds in the drug discovery pipeline. Figure 5.6 shows those four topics in one graph.

While industry analysts discuss issues on expectations and economic growth, they also translate, interpret, and evaluate scientific findings to a more general audience. The 200-topics model also yields clusters on particular research strategies, molecules, and treatments. Figure 5.7 gives an overview and Table 5.3 shows the terms of selected topics.

The spike in the topic "MMP inhibitors" matches the qualitatively gained insights on the increased expectations about their potential (in form of Marimastat and Batimastat) between 1995 and 1999. Larger in prominence, though, than "MMP inhibitors" are the topics "targeted therapies" and "mAbs" during the 1990s. In the topic "targeted therapies," terms such as "tumor, solid, cancer, targeting" cluster unusually often together. With new developments in the mid-1990s, this topic peaks for the first time, another peak is during the time of Herceptin's positive test results and approval, and a third peak is after 2002. Thus, in this corpus of industry analysts' reports, *targeted therapies* as a category can also be found.

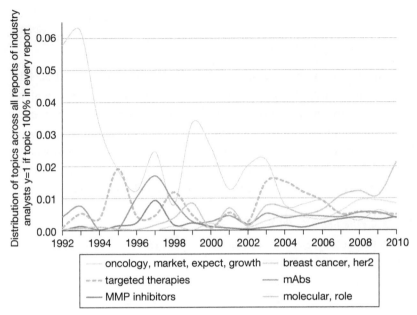

Figure 5.7 Selected Topics on Targets and Mechanism, Industry Analysts' Newsletters, 1992–2010

Table 5.3. 25 Top Terms of Nine Selected Topics, Industry Analysts' Newsletters, 1992–2010

oncology, market, expect, growth: market, markets, drug, drugs, report, companies, pharmaceutical, will, growth, industry, development, products, more, analysis, major, billion, key, reports, global, pipeline, also, oncology, therapies, expect, product

FDA approval: fda, unitedstates, approval, drug, has, regulatory, approved, application, european, company, review, products, announced, agency, product, orphan, been, development, marketing, use, submitted, actions, will, europe, granted

company, stock prices: company, stock, shares, common, will, million, its, inc, share, nasdaq, price, offering, securities, exchange, transaction, announced, purchase, approximately, per, closing, warrants, shareholders, capital, article, proceeds

drug discovery: drug, development, discovery, pharmaceutical, company, therapeutics, novel, inc, compounds, drugs, has, oncology, candidates, diseases, therapeutic, its, preclinical, molecule, targets, technology, small, lead, platform, potential, disease

breast cancer, HER2: breastcancer, breast, women, carcinoma, health, oncology, her2, trastuzumab, also, her-2, see, invasive, positive, via, neu, mda-mb-231, breastcancers, her2-positive, reports, tumor, receptor, mcf-7, tumors, herceptin, metastatic

targeted therapies: tumor, tumors, solid, cancers, blood, growth, normal, which, therapeutic, targeted, target, types, potential, these, metastatic, therapies, tissue, vessels, oncology, targeting, breast, carcinomas, malignant, may, drug

mAbs: antibody, antibodies, monoclonal, biotechnology, human, against, device, mab, antigen, medical, immunology, binding, immunotherapy, therapeutic, humanized, targeting, mabs, specific, surface, target, igg, using, recombinant, antigens, receptor

MMP inhibitors: matrix, invasion, expression, mmp-9, metastasis, protease, tumor, metalloproteinase, activity, extracellular, mmp-2, inhibitor, plasminogen, enzyme, tissue, metalloproteinases, human, activator, upa, mmp, protein, migration, proteins, inhibitors, urokinase

molecular, role: molecular, role, signaling, response, cellular, development, biology, these, pathways, mechanisms, been, findings, pathway, has, proteins, novel, may, published, growth, important, target, function, receptor, therapeutic, tumor

"Targeted therapies" as a topic is closely followed by the topic I have labeled "mAbs," clustering together terms such as "antibodies, monoclonal, recombinant." This topic peaks in 1997, when expectations into mAbs were at their height. The topic "molecular, role," which clusters together terms such as "development, pathways, mechanisms," focuses on biochemical mechanisms of molecular treatment. With growing scientific knowledge toward the end of the 2000s, it is a topic that is increasing as the time period analyzed ends.

Topics in Wire Reports, 1992–2010

Another text corpus of interest when studying the emergence of a market is that of companies' own press statements. The third large-scale data set contains wire reports and press statements of companies worldwide. The purpose of such reports is to publish some news about the company and to bring information about new developments into the public domain. Yet these reports are not merely informative; rather, as already evidenced in the earlier, qualitative analysis, they can also be evaluative and in particular capture beliefs, anticipations, and future expectations (see also Fortun 2001; McLaren-Hankin 2008; Tutton 2011).

As in the previous data sets, only a few articles could be found using the search string (biotech* AND cancer*) during the early 1990s. In 2000, 580 wire reports and press releases could be found, though (see Figure 5.8 for the distribution of documents over twenty-two years).[7] A total of n=12,672 articles were analyzed between 1992 and 2010.

Just as in the reports from the industry analysts, wire reports and press statements are filled with a wide range of topics: company management developments, the announcement of conferences and presentations, particular treatment types, specifics of diagnostics, and failed products. However, in contrast to reports from industry analysts, these reports consist of announcements, in which PR professionals package information.

Figure 5.9 shows selected topics on treatments and biochemical mechanisms. As becomes evident, the topic "marimastat, MMP" has the biggest peak, indeed of all topics, in 1995, when British Biotech was excited to announce progress and test results. This topic has practically disappeared after 2003. The topic "mAbs, platform" shows a peak and increase in the mid-2000s. As the top terms of the selected topics show (Table 5.4), this topic concerns technological developments of mAb discovery and manufacturing (platform technologies), which were taking off at that time and in which several companies were vying for attention. The topic "kinase inhibitors" is increasing throughout the period

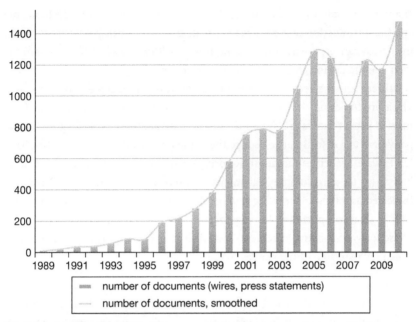

Figure 5.8 Distribution of Wire Reports and Press Statements 1989–2010, *PR Newswire* and *Business Wire*

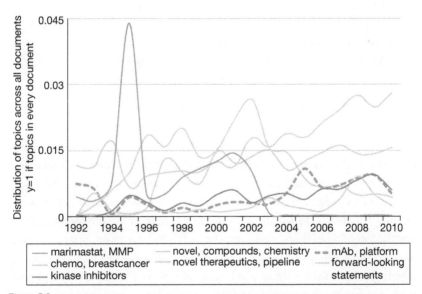

Figure 5.9 Selected Topics on Treatments and Biochemical Mechanisms, Wire Services, 1992–2010

analyzed. A topic I labeled "chemo, breast cancer" captures the development and the introduction of *doxorubicin* (Doxil) as a chemotherapeutic agent treating breast cancer; the associated drug Procrit within the same topic is used to treat anemia during chemotherapeutic treatment.

Three further selected topics from the 200-topics model are also of interest: These are topics on novelties and forward-looking statements. One of the features of press releases are "forward-looking statements," in which a given company inserts a disclaimer to reduce the company's liability under the law. The US Private Securities Litigation Reform Act of 1995 provides a so-called safe harbor for forward-looking statements in companies' press releases. Once it is indicated as a forward-looking statement, a company will not be held responsible if what is reported in the press release does not materialize as expected.[8] These forward-looking statements vary little between companies. For illustration, they read something similar to the following:

> This press release contains forward-looking statements that are subject to risks and uncertainties that could cause actual results to differ materially from those set forth in the forward-looking statements, including whether company XYZ becomes a major presence in the worldwide market for oncology and other pharmaceutical products, whether company XYZ can capitalize on the global demand for high-quality, low-cost pharmaceuticals, and the risk factors set forth in company XYZ's recent proxy statement filed with the Securities and Exchange Commission. These forward-looking statements represent company XYZ's judgment as of the date of this press release. Company XYZ disclaims any intent or obligation to update these forward-looking statements.

When looking at the distribution of topics over time (Figure 5.10), the topic "forward-looking statement" does not show great fluctuations beginning with its existence in 1996. In 1999 and in 2004, fewer wire reports in the corpus used disclaimers. The 25 top terms of the topic include "uncertainties, risks, may, future, materially" and capture the essence of a typical statement (see Table 5.4).

Two other topics appear noteworthy: the topic I have labeled "novel, compounds, chemistry," in which terms on drug discovery are grouped together, and the topic "novel therapeutics, pipeline," in which terms on pharmaceutical treatments are grouped together. The former thus refers to a topic in reports, which speaks to the chemical and biological side of drug discovery and production, while the latter topic points to potential therapeutics and novel drugs in clinical development. Both topics share several terms, such as "compound,

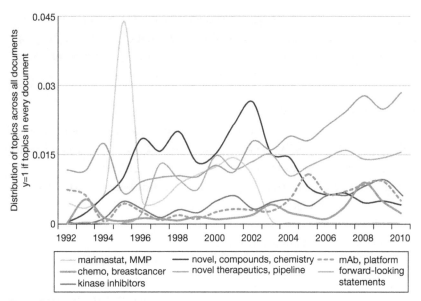

Figure 5.10 Selected Topics on Novelties and Forward-Looking Statements, Wire Services, 1992–2010

Table 5.4. 25 Top Terms of Seven Selected Topics, Wire Services 1992–2010

kinase inhibitors: kinase, inhibitors, inhibitor, cell, entremed, angiogenesis, small, protein, kinases, molecule, pathway, preclinical, growth, apoptosis, pathways, novel, clinicaltrials, signaling, cells, receptor, inhibition, activity, tumor, arqule, target

Marimastat, MMP: britishbiotech, marimastat, matrix, studies, program, clinicaltrials, study, inhibitors, batimastat, bresagen, treatment, collaboration, events, uncertainties, phase1, cell, inhibitor, e21r, bb-10153, strategy, forward-looking, regulatory, lung, rights, metalloenzyme

mAbs, platform: antibody, antibodies, monoclonal, therapeutic, biowa, potelligent, therapeutics, platform, human, aphton, adcc, mabs, cells, abgenix, more, protein, activity, targets, diseases, kyowa, cell, immune, igeneon, target, xencor

chemo, breast cancer: doxil, should, therapy, ortho, treatment, procrit, percent, reactions, may, information, not, severe, events, doxorubicin, have, who, receiving, dose, caelyx, use, most, serious, adverse, infusion, breast

novel compounds, chemistry: discovery, chemistry, targets, screening, compounds, novel, pharmaceutical, technologies, proprietary, target, pharmaceuticals, lead, platform, candidates, collaboration, drugs, small, companies, chemical, design, biology, molecule, programs, medicinal, molecular

novel therapeutics, pipeline: therapeutics, diseases, novel, clinical, drugs, pharmaceuticals, treatment, therapeutic, pipeline, discovery, candidates, developing, disease, lead, biopharmaceutical, proprietary, compounds, programs, small, potential, focused, platform, treat, more, molecule

forward-looking statements: forwardlooking, release, securities, risks, such, uncertainties, may, any, not, actual, these, differ, act, future, those, materially, could, factors, press, exchange, events, information, including, expectations, cause discovery, novel, drugs," though they vary in the importance of terms within each topic. The term "target" is part of the "novel compounds" topic. This topic is strongest between 1996 and 2003. Especially after 2003, the topic on "novel therapeutics, pipeline" is prominent in the corpus of wire reports.

Topics in Newspaper Articles, 1989–2010

Another macroscopic view offers the fourth large-scale data set of newspaper articles from the *Financial Times* and the *New York Times* between 1989 and 2010. Using the same search string of (biotech* and cancer*), the search and available data access yielded n=1,073 articles.[9] The distribution of articles across the entire period studied varies between 10 in 1991 and 129 in 2000. After many articles between 1995 and 1998, only 19 articles were published in 1999. Figure 5.11 displays the distribution over time.

Given the total number of documents, a 50-topic solution proved robust and interpretable. Among others, these topics include investments, patents, book reviews, particular organizations and companies, and treatment strategies. The topics concern economic evaluations and point to prominent companies

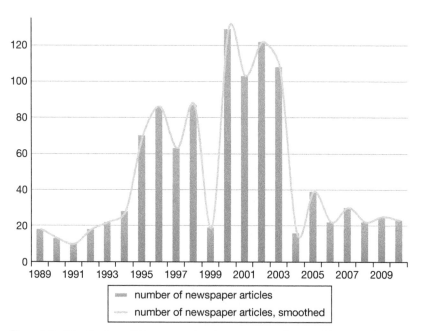

Figure 5.11 Distribution of Newspaper Articles 1989–2010, *Financial Times* and *New York Times*

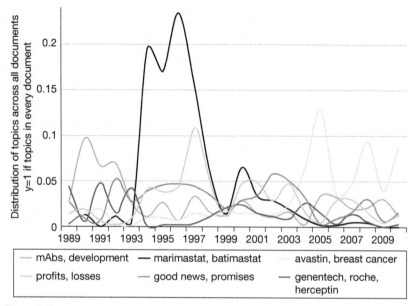

Figure 5.12 Selected Topics on Treatments and Developments, Newspapers, 1989–2010

journalists report on. At the same time, there are also human-interest stories of latest research developments packaged in a larger report.

Figure 5.12 shows selected topics on treatment and developments. Table 5.5 lists the selected topics and their 25 top terms. A first topic concerns the development of monoclonal antibodies in the early 1990s. The biggest peak in the entire corpus, however, with almost 25% of all articles making up that topic, is once more a topic concerning Marimastat and Batimastat in the mid-1990s. Here terms of "clinicaltrials, results, shares, failure, promising, potential" echo what the earlier qualitative analysis also found: a mixture of promises and failures, of value created at the stock market due to expectations, as well as falling stock prices due to disappointing trial results. After a momentary reappearance in 2000, when Marimastat's last clinical trial results showed no effect in cancer patients, this topic then disappears. A topic that rises in the mid-2000s instead is a topic on "Avastin and breast cancer." This topic refers to the results of clinical trials of Genentech's *bevacizumab* together with chemotherapy also for breast cancer. The rise of this topic captures approval for colon cancer (FDA 2004 and EMA 2005) and the provisional approval for breast cancer in 2008. The increase at the end of the period studied concerns media reports questioning the drugs' costs, effectiveness, and side effects. As indicated in Chapter 4,

Table 5.5. 25 Top Terms of Six Selected Topics, Newspapers, 1989–2010

mAbs, development: antibodies, immune, disease, human, antibody, diseases, body, monoclonal, mouse, blood, protein, developed, mice, clinicaltrial, hepatitis, proteins, using, treatments, develop, known, scientists, autoimmune, insulin, target, molecular

Marimastat, Batimastat: trials, marimastat, clinicaltrial, results, yesterday, share, progress, product, treatment, failure, britishbiotech, rights, price, cash, fell, regulators, value, final, batimastat, cancers, chiroscience, finance, news, promising, potential

Avastin, breast cancer: avastin, clinicaltrial, percent, treatment, trial, chemotherapy, genentech, trials, tumors, breast, approved, months, approval, effects, lung, data, results, cancers, colon, doctors, price, patient, study, tumor, expected

profits, losses: share, profits, pre-tax, expected, growth, quarter, earnings, loss, profit, rose, results, analysts, losses, half, net, reported, increased, rise, dividend, forecast, increase, compared, months, turnover, operating

good news, promises: sector, share, news, value, good, potential, price, high, just, past, way, future, financial, number, months, success, times, recent, promising, despite, long, confidence, early, analysts, time

Genentech, Roche, Herceptin: genentech, growth, big, business, roche, pharma, levinson, pharmaceutical, breast, companys, science, success, early, make, insulin, antibodies, herceptin, rituxan, medicines, hormone, francisco, proteins, monoclonal, heart, swiss

in 2011 Avastin's approval for the treatment of breast cancer was revoked in the United States.

Other topics concern economic evaluation and individual companies. Figure 5.13 emphasizes topics, which group together terms on "profits, losses," "good news, promises," and on "Genentech, Roche, Herceptin." The topic on "profits, losses" peaks in 1997, at a time when the market of innovative breast cancer therapeutics was developing but highly ambiguous. The topic "good news, promises" is evident during the Batimastat/Marimastat years and again in the early 2000s, when journalists discuss potentially successful drugs. The topic "Genentech, Roche, Herceptin" captures terms of a successful monoclonal antibodies' treatment. By 2004, Herceptin had become the gold standard against which other treatments are evaluated in terms of targeted biochemical performance and economic performance.

These topic model analyses of four different textual corpora charted global developments and provided a macroscopic view over twenty-two years of developments. The analyses thus went beyond select samples of texts or particular approved products to reveal the entire discussion in different fields collaboratively constructing the market of breast cancer therapeutics.

Substantively and most importantly, the analyses showed a shift from a chemotherapeutic topic to one of new, targeted approaches. "Targeted therapies"

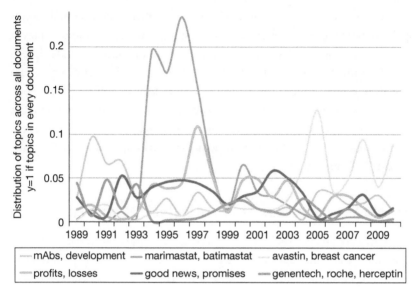

Figure 5.13 Selected Topics on Economic Evaluation and Genentech, Newspapers, 1989–2010

came to form a category beginning with 2002. The shift from chemo to targeted approaches was most pronounced in the scientific discussions. The role of targeted therapies was a topic in the industry analysts' reports as well, beginning with 2003. In wire reports, the topic of novel therapeutics, capturing novel molecular drugs, began to rise especially in 2003.

Second, the topic model analyses gave insights into the trajectories of biochemical mechanisms, molecules, and compounds from different perspectives. From the scientific, oncological discussions, specific shifts from interleukins to immunotherapy and mTOR pathway became evident. MMP inhibitors, including the failed product of the late 1990s Marimastat and Batimastat, were also topical in scientific abstracts, industry analysts' reports, wire reports, and newspaper reports. The formerly failed research strategy regained prominence in the mid-2000s.

Third, the analyses highlighted topics on economic evaluation. Industry analysts had high expectations for oncology products in the early 1990s and again in the early 2000s, once Herceptin was approved. These expectations of a growing market were disappointed when no new products received approval until 2007.

While it was instructive to analyze these shifts in discussion based on large textual corpora, the topic model algorithm used does not allow exploring the *structure* of co-occurring terms within a topic and then the dynamic of that structure over time. As noted in the beginning of this section, several caveats apply to topic modeling, regarding the substantive interpretation of topics as they are based on a probabilistic distribution of terms across all texts, as well as to the number of topics. Here topic modeling was used heuristically to gain insights into select trajectories of categories over time in four textual corpora, while acknowledging the methodological limitations. The next section picks up on these limitations and delves further into categories when it analyzes co-occurrence of terms based on network analytic techniques to show the shifts and drifts of category emergence.

Semantic Network Analyses: Shifts and Drifts of Category Emergence

Using network analytic tools and visualization techniques, research has shown communities and the "backbone" of particular scientific fields (e.g., Boyack et al. 2005; Cambrosio et al. 2006; Moody 2004). These works use the idea of duality discussed as the "intersection of social circles" by Georg Simmel (1992 [1908]). Translated into network analytic form by Breiger (1974), the idea is to bring persons and groups they belong to into an affiliation matrix of persons by groups. Using matrix multiplication, i.e., multiplying the matrix by its transpose, we can identify membership overlap of persons in groups and membership overlap of groups in persons. Persons are linked by the groups they belong to. Groups are linked by the persons they share. Affiliation matrices, grounded theoretically in the idea of duality, have shaped network analytic thinking and formal network analysis ever since.[10]

Analyses of citations and co-citations also rely on these principles. Such works have mapped clusters of membership of authors/journals/keywords based on their affiliation with each other. Typically, these two types of relations are visualized in the same image. For example, using tools of natural language processing to extract terms as well as research organizations from the textual corpus of 30,139 scientific abstracts, an affiliation network can be mapped. Research organizations are affiliated with particular terms, e.g., the University of Heidelberg near the top of the visualization with "tumor cells" and "multivariate analysis" (Figure 5.14).

This section uses these insights and carries them further when it considers the co-occurrence patterns of terms over time for the data set on oncological, scientific discussions. Now semantic network analytic techniques are used, first, to show shifts and drifts in category formation over time, and second, to elaborate on the focus category *targeted therapies*, which features prominently

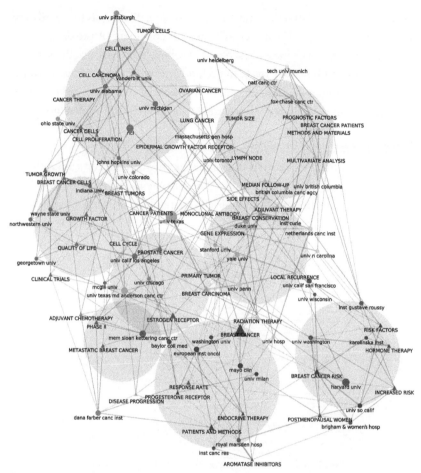

Figure 5.14 The Field of Breast Cancer Therapeutics: Research Organizations and Their Associated Terms Based on Abstract Extraction, Estimated and Visualized Using the CorText Platform, 1990–2010

in the scientific discussions, in a way that was not possible with the topic modeling techniques used.

The analyses are based on frequently co-occurring terms, including multiword phrases such as "metastatic breast cancer" or "radiation therapy," which I extracted from the data set of 30,139 oncological abstracts using natural language processing (NLP) techniques (e.g., Jurafsky/Martin 2009). In the following analyses I use tools developed as part of the CorText platform (see also Chavalarias/Cointet 2013; Rule et al. 2015; Venturini et al. 2014).[11] After preparatory steps of extraction and NLP, a first calculative step is to identify the co-occurrence of term "i" with term "j" in each document. A proximity score then measures the relatedness of each pair of terms. In the visualizations, terms are the nodes and how they relate to each other are the edges of the network. The Louvain community detection algorithm (Blondel et al. 2008) groups terms together into cohesive clusters, which are then marked by different colors.

Categories of Research Strategies

The analysis of the entire corpus of documents (n=30,139 documents with abstracts) yields a semantic map of oncological discussions from 1990 to 2010. Figure 5.15 is a visual representation of the eleven master categories of twenty-one

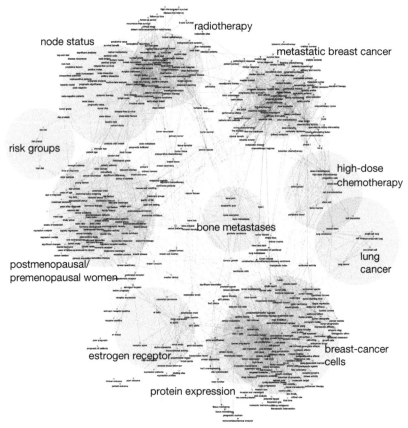

Figure 5.15 The Global Semantic Network Structure of Oncological Discussions on Breast Cancer Therapy, Extracts from Web of Science Abstracts, 1990–2010

years of breast cancer therapy discussions based on a 1000x1000 extracted multiterms matrix and their co-relations. Visible here are 488 nodes and 5,164 edges between them. Categories are labeled based on the lexical content of each cluster as established by the Louvain algorithm. Both the internal structure of the clusters and their position in relation to others are meaningful. Starting on the upper right "metastatic breast cancer" is a category that summarizes a state in the disease; its nodes are connected to other clusters concerning cancer treatments. "Radiotherapy" and "high-dose chemotherapy" are the labels of such treatment categories. Nodes on cell stem in "high-dose chemotherapy" also overlap with the category "lung cancer." "Bone metastases" are located in the middle of the network map. Yet "metastatic breast cancer" is also multiply connected to the periodization category "postmenopausal/premenopausal women." "Risk groups" and "node status" can be found on the upper left hand. "Node status" connects terms on the state of cancer nodes and thus sits between "risk groups" and "radiotherapy" and its nodes on local recurrence. The periodization cluster "postmenopausal/premenopausal women" unites several aspects, including weight gain and quality of life, but also hormonal treatment types. This cluster then links also to "estrogen receptor," which bridges via the node gene expression to a more technical cluster on "protein expression." This cluster summarizes the technical advances necessary for molecular treatments. At the bottom, "breast-cancer cells" clusters together all terms related to cancer cells: targets, genes, mechanisms.

However, just as in the topic model analyses, my interest lies in detecting possible shifts in research strategies on breast cancer therapeutics. I am not interested in analyzing *all* the possible themes the corpus entails. I thus took another step and extracted the keywords that the authors included with their abstracts (all documents with abstracts also have keywords). Though this is possibly less detailed than the texts of the abstracts, it allows getting at the gist of the abstracts' contribution.

When taking the top 100 keywords and their 498 co-relations from the entire period 1990–2010 into account, the semantic network shows seven different clusters (Figure 5.16).[12] The labels are based on the lexical content of the cluster. They refer to the main categories of oncological discussions over twenty-one years as expressed in the keywords of articles. The arrangement of all clusters as well as the nodes making up the clusters are the result of calculation. The categories mapping, starting on the upper right side, are "quality of life," "apoptosis," "targeted therapy," "chemotherapy," "adjuvant chemotherapy,"

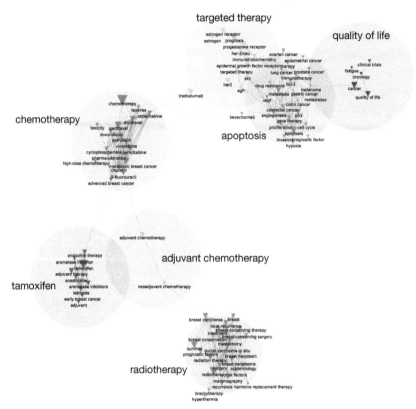

Figure 5.16 The Global Semantic Network Structure of Oncological Discussions on Breast Cancer Therapy, Extracted Keywords, Web of Science, 1990–2010

"tamoxifen," and "radiotherapy." "Apoptosis" overlaps with "quality of life" and "targeted therapy." Note that the nodes trastuzumab and bevacizumab link the clusters "chemotherapy" with "targeted therapy" and "apoptosis," respectively. They serve as bridges. "Adjuvant chemotherapy" used in combination with hormonal treatment overlaps with "tamoxifen."

How did the categories change over time? When looking at the same data set in five separate time periods, shifts in category formation become traceable. Note that these categories are formed from the information given by authors when the article was written. For purposes of visual readability, I use the top 50 keywords in each time period and map their co-relations (Figure 5.17). The first period consists of the early years and ranges five years (1990–1994, P1), the following four time periods consist of four years each: 1995–1998 (P2) is the period until the approval of Herceptin; 1999–2002 (P3) is the immediate

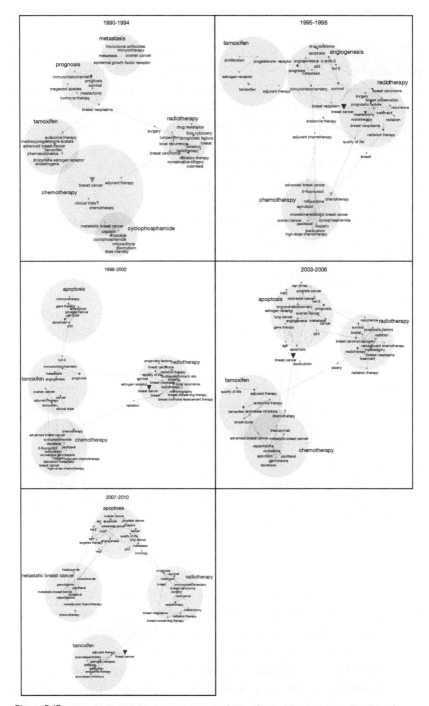

Figure 5.17 Discourse Categories over Time of Oncological Discussions, Extracted Keywords, Web of Science, 1990–2010, Five Time Periods

years after Herceptin's approval; 2003–2006 (P4) is a period of disappointment in terms of product approval; 2007–2010 (P5) marks the last phase with new products approved. The biggest shift is obvious from the first to the last period. It is possible that category labels remain the same, while the terms making up the category shift. For instance, while chemotherapeutic treatments change over time, though the category continues to exist, it is only when newly developed molecular treatments are introduced to be used in tandem with chemotherapy that the category disappears.

The first period (1990–1994) captures the early years of innovative cancer therapeutics, when monoclonal antibodies were a novelty and when hormonal and chemotherapeutic approaches as well as radiotherapy were typical. The second period (1995–1998) represents the time when new research approaches, such as focusing on angiogenesis, began. The third period (1999–2002) covers the early years after Herceptin's approval. No nodes related to Herceptin are clustered during that phase, instead we find the research strategies of apoptosis, gene therapy, and immunotherapy (first cluster on top on "apoptosis"). In the fourth period (2003–2006), HER2 is part of the "apoptosis" cluster, in which the research strategy angiogenesis is also located. Clusters "tamoxifen" and "chemotherapy" overlap at a distance from "apoptosis." In the last time period (2007–2010), the "apoptosis" cluster bridges between the clusters "metastatic breast cancer" and "radiation therapy"; among others, mTOR, HER2, and VEGF are clustered together, while the molecular products trastuzumab and bevacizumab are clustered together with the chemotherapeutic agents used in regimen tandem in the cluster "metastatic breast cancer." Here we have thus seen drifts and shifts in category formation.

Dynamics of Category Targeted Therapies

Analyses of keywords over time offered insights into the larger categories, with "targeted therapy" being visible as a category in the global network structure (Figure 5.16). Beginning with 2003, HER2 was a keyword node, and it remained a node, bridging between categories "apoptosis" and "metastatic breast cancer" in the mapping of the top 50 keywords in 2007–2010.

As shown earlier from qualitative insights, the category *targeted therapies* went through different phases of labeling, ignoring, co-opting, before becoming institutionalized. Also, results from topic model analyses showed that categories of "novel approaches" and "targeted therapies" began to rise in 2002.

Patterns in Meaning-Making 117

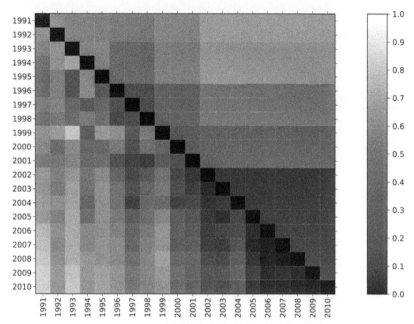

Figure 5.18 Transition Matrix Comparing Each Year How "Target" Is Used and Associated in Oncological Discussions. A shift to similarity becomes evident beginning in 2002.

Within the oncological discussions, what were the different understandings of "targeted therapy" over the larger time frame?

I now inquire further into the dynamics of what would emerge as *targeted therapy* and as such would help to form a market. For that I need to consider the context of meaning first and foremost of the term "target." I conducted a lexical extraction of noun phrases with maximum length of three from all Web of Science abstracts available with the pivot term "target" at the sentence level. In each extracted term, then, "target" appears.

Comparing the similarity of oncological scientific discussions on "target" across all years using a transition matrix that calculated five time periods, I find that the discussion is dissimilar until 2002. Terms in abstracts in 1999 were particularly different from the other years. Beginning with 2002, evidently the understanding of the term "target" becomes much more similar to that of the future years. Figure 5.18 shows the transition matrix. It uses a 1000x1000 top term matrix from the extracted data set on "target." Each cell in the matrix compares the term vectors for two years of abstracts, showing their dissimilarity on a scale from 0 (most similar) to 1 (most dissimilar).

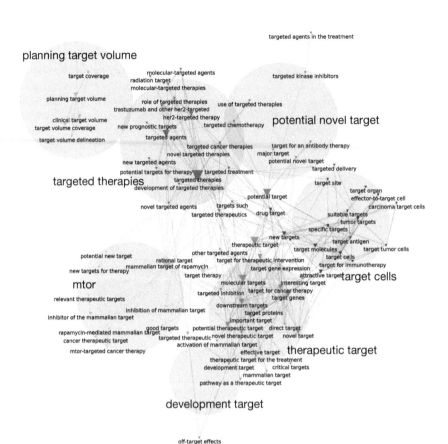

Figure 5.19 The Global Semantic Network Structure of Discussions on "Target" as Extracted from All Oncological Abstracts on Breast Cancer Therapy, Web of Science, 1990–2010

Darker regions represent areas of substantial similarity, with black being the most similar. Light gray cells represent years in which the abstracts departed from that of other years. Abstracts in 1999 were most dissimilar to those in 1993, and abstracts in 2010 were also very dissimilar from those in 1991, though very similar to those in 2009. The triangle on the upper right-hand side of the matrix presents the calculated similarities and dissimilarities between five periods.

Figure 5.19 then shows the global semantic network on discussions of "target" as extracted from all oncological abstracts on breast cancer therapy, 1990–2010. Based on 85 nodes and 578 edges, the Louvain community detection

Patterns in Meaning-Making 119

algorithm finds seven clusters, of which six clusters are connected to each other, while "planning target volume" is unconnected. On top of the mapped network, the cluster "targeted therapies" is located. It assembles nodes on HER2, novel targeted therapies, and novel targeted agents. It overlaps and shares nodes with the cluster on "targeted chemotherapy." The "targeted therapies" cluster is also connected to the other four clusters: "target cells," "therapeutic target," "development target," and "mtor." "Target cells" connects terms of different target types. "Therapeutic targets" connects terms on molecular and cancer targets. Related to the latter is "development target." In cluster "mtor" terms connect that relate to the mammalian target of rapamycin and its inhibitors.

The transition matrix indicated that shifts and drifts occurred in how the term "target" was used in the scientific discussions. The global network structure showed that seven categories exist from 1990 to 2010. To obtain a better substantive understanding of the shifts and drifts that are the foundation for such global map, I look at how "target" was used in different time periods. I used the periodization as above (P1: 1990–1994; P2: 1995–1998; P3: 1999–2002; P4: 2003–2006; P5: 2007–2010) to detect changes.

Figure 5.20 shows the results. In each of the periods P1 and P2 the Louvain algorithm finds two clusters. "Target antigen" continues from P1 to P2, while the cluster on "target tumor cells" disappears. Instead, an unconnected cluster with the lexically based label "potential target" is found. It unites multiterms such as "targets for immunotherapy" and "target molecules." In P3 (1999–2002), then, the number of clusters increases to five. The field has apparently changed. We see here multiple ideas connected to the term "target" in the scientific discussions. The former cluster "target antigen" now primarily is about "therapeutic target." Linked to it are all other clusters: "mtor" and "potential target," the latter of which is an extension of the "potential target" cluster of P2, and "molecular targets" and "rational target," which cluster together nodes of new news and ideas. The term targeted therapies is a node connecting clusters "rational target," "molecular targets," and "therapeutic target." The five-cluster solution develops even further in P4 (2003–2006), when the cluster "therapeutic target" continues to be located in between clusters "targeted therapies" (with overlap), "target site," "target cells," and "mtor." An unconnected cluster on "planning target volume" has developed, which links together terms on dosage of treatment for patients. The cluster "mtor," in overlap with the cluster "target cells," links together terms on new and novel targets. This echoes the trajectories detected earlier using topic modeling. The cluster "targeted therapies," with its overlap to

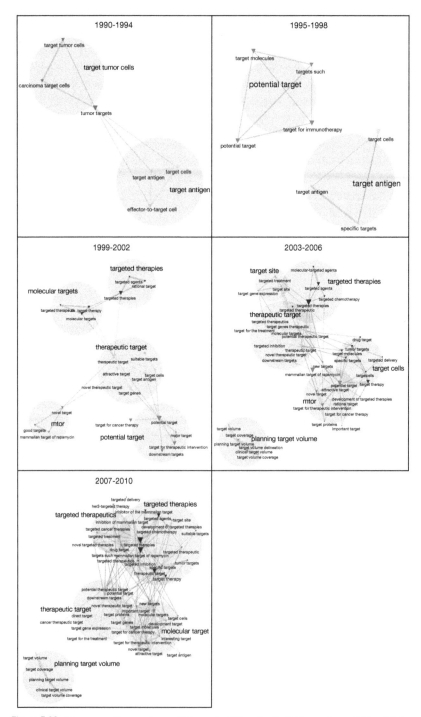

Figure 5.20 Discourse Categories on "Target" of Oncological Discussions, Web of Science, 1990–2010, Five Time Periods

"therapeutic target" and "target site," appears as an established category. In P5 (2007–2010) the category "targeted therapies" is even larger than before: it has now merged with some nodes of the cluster "target cells" from the former period. It now also includes the node mammalian target of rapamycin (mTOR), which at that time was widely discussed as a decisive for inhibiting cancer growth (see also topic trajectories, e.g., in Figure 5.4). Adjoining is the cluster "targeted therapeutics," which unites nodes on treatments and therapies as general concepts, e.g., HER2-targeted therapy. The cluster "therapeutic target" continues to exist, now with an increased focus on target genes. It is located adjacent to the cluster "molecular target," which unites particular discussions e.g., on attractive target or interesting target. The cluster "planning target volume" again sits separate from all other categories, as it pertains to another discussion.

Based on these five periods of semantic network clusters, a Sankey diagram can be built (Figure 5.21, part of CorText platform). It shows the flows of categories over time and highlight consistencies as well as shifts and drifts. Some categories emerge; others merge and fold. Again, while a category may remain the same from one period to the next, the terms making up the category might change. The five-period solution finds one category as existing throughout the twenty-one years analyzed: this is the category that starts out as "target antigen," then shifts to "therapeutic target" (at bottom of Figure 5.21). It starts out

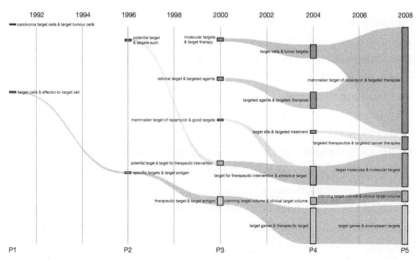

Figure 5.21 Sankey Diagram, Showing the Flows and Drifts between Different "Target" Categories over Time, 1990–2010 (Five Periods Marked at Bottom)

with co-occurrences on "effector-to-target cells" and ends with terms on "target genes," both belonging to the same category. All categories swell in content of co-occurrences at the end of the time studied. The second longest existing category starts in P2 and concerns the first the category on "potential target." By P4 it has merged with "mTOR," highlighting that these targets are no longer only potential and attractive, but rather are "molecular targets." The Sankey diagram also shows how in P3 (1999–2002) the field of discussion shifted to more categories. The two categories from P3 merge into one category on "targeted therapies" in the last period studied. We also see two categories arising in P4: "planning target volume" and "targeted therapeutics." Also note that one category from P1 on "target tumor cells" remains unconnected to the rest of the scientific discussion.

The central finding of the semantic network analyses is that indeed the category *targeted therapies* emerged beginning with 2002. Using the scientific discussions, analyses of co-occurrences of terms over time show the role this category played for holding the scientific discussion together, e.g., when particular compounds were bridging between categories. It also highlighted how pivotal the scientific discussions on novel, targeted therapies were for moving the field of science forward. At the end of the analyzed period of time, *targeted therapies* was an established and well-institutionalized category, with repercussions, as the analyses of other corpora showed, in financial and industry analysis.

These large-scale textual analyses confirm and support the insights gained from the prior qualitative inquiry into different phases, including the trading zone phase. The term "target" was used in multiple meanings, including "targeted chemotherapy," expectantly as "potential target" but also as "novel target," evaluating an existent state of research. In addition to finding different categories that capture the entire scientific discussion on breast cancer therapies over twenty-one years from the perspective of when the authors of each scientific article were writing it, using extracted terms (Figure 5.15) as well as keywords (Figure 5.16), the analyses also provided insights into the temporal dynamics of category formations. By way of patterned temporariness, the analyses indicated shifts and drifts from one co-occurring cluster of terms to another. While individual terms may have obtained meaning in another set of co-occurring terms when they switched, clusters nevertheless maintained a master category identity.

For one, this chapter has shown the utility of large-scale text analytic methods. Text analytic methods allow for macroscopic view onto entire fields that would otherwise be unavailable. The analyses are able to capture developments of twenty-one years in complexity-reduced visualizations. While topic modeling presents a ranking of terms as a topic, the network-based text analytical approach uses the idea of co-occurrence also for finding community structure and their lexical title. Such large-scale textual analyses of "old big data" thus provide insights into the dynamics of an emergent formation as they allow taking the *in situ* sensemaking into account.

Moreover, this chapter has shown the dynamics of the construction and then institutionalization of *targeted therapies* as a category in the emergent market across different, though interconnected, fields. Both topic model analyses and semantic network analyses showed that *targeted therapies* became a prominent category in the early 2000s, developing into the most important category at the end of the 2010. The analyses have also highlighted developments of scientific findings alongside stories of industry analysts, companies, and journalists. The findings support previous results from the qualitative analyses of stories dominated by expectations about a future market in the early 1990s. With the unfulfilled expectations of British Biotech's products, topical discussions on profits wane. Beginning with the 2000s, then, topical discussions shift to drug discovery, stock prices, and a new round of stories of expectations and promises along with the discussion of novel compounds and therapeutics in the pipeline. By the mid-2000s, the field composed of the interrelated network domains of science, commerce, industry, and journalism had collaboratively constructed a new category, which laid the grounds to induce a shift in market structure. At that time, new products and new approaches with newly raised expectations became evident from the analyses, e.g., discussions on monoclonal antibodies or on immunotherapy. As the macroscopic analyses of the emergent category *targeted therapies* highlight, through their stories, a diverse set of involved actors contest, negotiate, and at least temporarily settle on an agreement of what the emergent field of research and market is about. Together, with their stories, they categorize and make sense.

Conclusion
Markets from Stories

THROUGH THE EMPIRICAL STUDY of innovative breast cancer therapeutics, this book has challenged the view of markets as already existing forums of exchange of goods. It asked how actions are coordinated before a market exists, i.e., in situations when it is unclear who participates, who will profit, and which product may be valuable. The perspective offered suggests understanding markets as networks of sociocognitive and sociomaterial actors that are based on stories. Examining textual data using qualitative and computational methods, the book has presented a cultural analysis of the emergence of a new social formation: the market for targeted breast cancer therapeutics.

In this last chapter, I highlight findings of the empirical inquiry and relate them to my analytic model and explanatory theory. I then indicate my contributions to ongoing discussion in the field of sociology and suggest how the approach that stories are the central mechanism for the emergence of markets applies more generally to the study of emerging social formations and point to areas of future research. I conclude with a look at developments in the field of cancer research, which continue to stir high hopes for millions of patients.

"Markets from Stories"

Throughout this book, the main argument has been that markets come from stories. Major contributions in economic sociology have focused on direct relational ties and their meaning for actors, on power relations, on the observation of competitors and status ordering, on sociotechnical arrangements, and on orders of worth to explain how markets are constituted. In addition, a prominent approach that focuses on the emergence of organizations and markets (Padgett/Powell 2012b) provides a multilayered structuralist model of social transformation. Yet such structuralist focus alone does not suffice to explain how actors position themselves vis-à-vis one another, or how they evaluate ideas or concepts; interpretation and evaluation are necessary to allow comparison. Analytical models of economic action that include the role of reflexive cognition, interpretation, and evaluation overcome the structuralist limitations (Stark 2009). My strategy was to analytically recombine those two models. Such combination yields a model of multiple network domains, which are interrelated, and which need to be studied over time, with a focus on interpretation and evaluation. Yet, missing from such a model, I have argued, is a crucial element that makes interpretation and evaluation possible: stories. Consequently, to the recombined analytical model I add a focus on stories, with a particular emphasis on public stories of the future.

In stories, actors state findings, present an analysis of the situation, interpret, evaluate, and voice their expectations about an unknown future. Their actions and their stories about these actions serve as signals to others. Each actor's stories cannot be understood in isolation, but rather must be understood as entangled with other actors and their stories, who in the studied case of an economic setting may be potential competitors, analysts and thus evaluators, financial investors, and other expert communities. Actors thus not only observe their competitors but also communicate. In communicating, these entangled actors position themselves vis-à-vis others. Actors may also narratively attach themselves to others, their ideas and research strategies, their expectations, and, as we have seen, even their molecules. From a network perspective, then, stories are the ties that connect actors, actions, including research strategies, interpretations, and expectations over time.

Actors use stories to grapple with the uncertainties and ambiguities of newness. Narrative constructions in general are fundamental to economic action (Beckert 2016). More specifically, actors tell stories of different temporalities, relating to the past, present, or future. Stories of the future play a special role for economic actions. With their imaginations and expectations about what the future holds, they provide orientation "*despite* the uncertainty inherent in the situation" (Beckert 2013: 222). They do so by offering "convincing" accounts of the future, which are comprehensible and plausible, thereby momentarily "overlooking" uncertainty. Because of this ability to temporarily suspend uncertainty, stories of the future are fundamental for economic actions. To be sure, regardless of whether stories of the future correctly predict the future, which can be seen as highly unlikely, they nevertheless influence action in the present.

Applying this theory to an empirical case, the book showed that stories are both primary data for analysis and, in an empirical setting, forces for contextual, situational meaning-making. Stories helped actors to establish an interpretation of the perceived social formation they deemed themselves entangled in. This formation began as a market of expectations during the initial search for innovation and before an actual product existed; over time it transformed when it settled on a new category, *targeted therapies*. Stories also helped actors to construct representations of others. Just as in other social settings, in economic settings too interpretation and meaning-making are essential for decision-making processes. The analyses showed that because of convincing stories, however temporary or long-lasting, investors make decisions, companies invest funds, and competitors may alter the line of research. Stories thus motivate, orient, and coordinate actions. At the same time, the involved actors from the network domains science, business, industry, and journalism relied on a heterogeneous set of actors, including regulatory agencies, molecules, clinical test results, investors, and scientific findings, against which they built their interpretations and evaluations. Thus, while stories should be understood as the microfoundations of decision-making and are necessary for economic action to take place, stories alone cannot make up a market; actors who interpret and evaluate them are needed.

The book showed that an analytical and empirical focus on stories allows for an inquiry into the relational constitution over time of actors, objects, and processes, providing descriptions and evaluation. The inquiry followed a basic tenet of both structuralist network analysis and pragmatist actor-network theory: there is no *a priori* ascription of who the powerful actors are. Through the field of biotech, a realm of constant innovation and financialization of innovation, the inquiry followed a diversity of actors in making associations. Positions of power got established in the processes of making connections, and at times were only of temporary relevance.

Moreover, the book also highlighted that multiple networks of actors tell stories. It is thus not one story that is told at any given time, but rather many stories of many actors vying to participate in defining what counts in a market. Subsequently, the perspective advanced here took a plurality of stories and perspectives and a plurality of connections into account—concomitantly and crisscrossing each actor. In these multiple stories, multiple evaluative principles were at play. Stories of the future made promises and pretended to know what the future will hold. We saw multiple and competing interpretations of these stories of the future. Actors narratively competed for "their story" and interpretation to be of worth to others. Stories imitated other stories ("we are comparable to others"); they also anti-imitated and dis-associated themselves from already existing stories ("we are like no one else"). In these competitive evaluation processes, meaning was temporarily shared, and actions were coordinated amid ambiguities.

The perspective the book advanced thus expanded prior approaches to the study of markets and economic stories when it studied multiple perspectives using a plurality of stories with multiple methods in order to detect entangled meaning-making processes. Such inquiry resulted in a cultural analysis of the emergence of a market. Grounded in what has been called relational sociology, the theoretical approach that combines insights from structuralist network analysis with US cultural sociology and considers networks as constituted by cultural processes of communicative interactions, the study has inquired into meaning-making processes based on both qualitative as well as formal analyses. The empirical inquiry focused on public texts of economic actors involved in the search for therapeutic treatments for breast cancer. Of primary interest were biotech companies as integral players in this search. As organizations, biotechs are at the intersection of different fields with varying principles of evaluation and what counts as worthy: science, business, and finance. With the addition of journalists as public interpreters and evaluators, participants and analysts of these three fields presented the focus of the empirical analyses.

The Emergence of a Market

Through foregrounding the explanatory theory of markets from stories and developing an analytical model, the empirical inquiry found two elements to be fundamental for the emergence of a new market: first, the temporary establishment of a *market of expectations* that provided structure for competing actors to compare themselves against and position themselves accordingly; and second, and after a first tested and approved product existed, a phase of contestations and confirmations about *categorizing* what this new market could be about.

At the time when no innovative breast cancer treatment product existed, biotechs, financial analysts, journalists, and scientists, through their stories of the future, collaboratively created, first, a market of expectations. The qualitative analyses showed that in addition to latest research results or funding successes, biotechs traded stories on expectations, projections, and imaginations on what the future held. Despite the ambiguity of how to evaluate new molecule, these stories provided orientation for the involved actors because they offered plausible accounts. Share prices, the analysis showed, reflected future expectations, while setbacks in research results damaged such expectations and sank stocks. Biotechs' stories of the future tied clinical results of the present to the scientific and also financial performance in the future. Yet it was not only biotechs who were able to momentarily distance themselves from uncertainties and ambiguities; this also held for journalists and financial analysts, who voiced their bets of the future when they interpreted and evaluated biotechs' research reports. In these evaluation processes, meaning was locally and temporarily shared. Hooking on to stories about the future full of expectations and promises helped the involved actors to cope with uncertainties of the situation while concomitantly strengthening their cognitive interdependence. Together and with their entangled stories, they coordinated and collaboratively created a market of expectations. In a first step, the market for targeted breast cancer therapies emerged not only because of timely scientific developments, but also because of interactions between companies and stories of financial and therapeutic expectations.

Furthermore, once this market of expectations had a first institutionally approved product (*trastuzumab*/Herceptin), the entangled sociocognitive network of scientists, analysts, evaluators, and competitors had a reference to compare against. In the next phase of market emergence, the involved actors were entangled in categorization processes when they evaluated and interpreted what they and others were about, where products fit into, which market they belonged to, how to interpret such market, and overall, what their new social formation as about. At first this was a process of contestations and disagreement about labels, before they collaborative came to agree, albeit temporarily, on a new category, *targeted therapies*, to which a diversity of actors could attach. The study found that this category formed out of multiple stories across actors as temporary bundles of meaning, in order to define what "it" is about and who others are when the setting is fraught with uncertainties and meanings are multiple. A category, it turns out, is collaboratively constructed and can be, though need not be, temporary. Before a category exists, there are labels that need to be filled with meaning. Such categories may experience rivalry from other attempts to label a bundle of stories. The analyses showed that categories need their own supporters in order to successfully stick. Furthermore, the category *targeted therapies* also signaled a *network folding*: practices of what it meant to do cancer treatment research changed, molecular biology skills were needed, pharmaceutical companies no longer expected a cure-all blockbuster drug. The concept of *targeted therapies* changed "the ways things are done" (Padgett/Powell 2012b: 5).

The macroscopic, large text analytic examination over twenty-one years found support for both phases, a market of expectations and categorization, in four interconnected network domains of science, industry, business, and journalism. Using computational methods to analyze large textual corpora, the analyses demonstrated how the category *targeted therapies* came to be institutionalized in all four network domains. First, using topic modeling, the analyses highlighted different field-specific trajectories. Discussions in the field of oncology allowed for a tracing of different research strategies for breast cancer therapies, which translated into business strategies for biotechs. Business analysts' reports showed ups and downs of expectations of the market. Biotechs themselves reported on their progress and successes and told about the role of novel targets and compounds. Journalists reported on and interpreted new molecules and promising treatments. Also, strikingly evident from these analyses was the high degree of expectations during the early phase of market emergence when a "cure for cancer" was projected onto MMP inhibitors. After their failure in clinical trials in the late 1990s and early 2000s, MMP inhibitors disappeared from all but the scientific corpora; yet the oncological research continued to inquire into their mechanism (Brower 1999; Coussens et al. 2002).[1] Using tools of semantic network analysis and focusing on the structures of term co-occurrences in the scientific corpus, the analyses then showed the shifts and drifts of research strategies over time. A focus on the term "target" examined its relations to other terms over the entire period of twenty-one years. The analysis could again show how *targeted therapies* became a category. The collaborative category construction across networks bolsters the emergence of the new market of targeted breast cancer therapeutics.

The empirical inquiries have shown that the formation of a new category through the coordination of stories is a social and cultural project that involves a diversity of actors. Moreover, it is a crucial mechanism of how new markets, and more generally, of how new social formations, come about.

In hindsight, Herceptin/*trastuzumab* is the treatment that together with its expectations, interpretations, and categorization was the first mover in leading toward several changes. It was one of the first drugs to be used in conjunction with a genetic test.[2] Indeed, *trastuzumab* has by now come to be regarded a first "tailor-made treatment" because it targeted a particular genetic makeup (Nathan 2007; Pusztai 2008). This idea of targeted treatment has become popularized in the notion of "personalized medicine" (e.g., Amir-Aslani/Mangematin 2010; Aspinall/Hamermesh 2007; Chin et al. 2011; Ellsworth et al. 2010; Hedgecoe 2004; Jørgensen/Winther 2009; Kho et al. 2010; Langreth/Waldholz 1999; Schilsky 2010), which is not personalized at the level of each individual, but rather at the level of patient groups showing particular genomic signatures in their tumors. Developments in "precision medicine" have pushed to further individualize treatments with biomarkers (Salgado et al. 2018).[3]

From an *ex post* perspective, then, *trastuzumab* was the harbinger of a change that has moved the field toward "personalized medicine," leading a wave of new drug developments using targeted therapeutics. *Trastuzumab* also needs to be reinterpreted as a first of a new type of *target-specific blockbuster* for its developers. Before it lost its patent, no other breast cancer treatment yielded higher revenues[4]; in 2014 *trastuzumab* ranked in global sales among the top ten pharmaceutical products, with sales of US$ 4.68 billion for its breast cancer indication. Thus, and as opposed to initial evaluations, *trastuzumab* is a blockbuster in terms of sales after all—and has been so since 2005 (Lawrence 2007a). Moreover, while helping to establish the market for breast cancer targeted therapies along with the other molecules and their stories told, *trastuzumab* also became part of a growing market of therapeutic monoclonal antibodies. These are not only the cornerstone of modern biotherapeutics, spanning across multiple therapeutic categories, but also constitute the best-selling pharmaceutical products (Ecker et al. 2015).

Yet, to reiterate a foundational argument of the book, *trastuzumab* is only one actor in a network of sociocognitive and sociomaterial actors that together came to form a market. The perspective has been one of multiple network domains and how they intercalate via stories dynamically over time. Stories of the future as expressed in business strategies, scientific test results, or financial recommendations are expectations that drive the construction of a market. I have developed an analytical model that shows that due to cognitive interdependence as well as, in this empirical case, tangible test results, first, a market of expectations exists. In a second step a diversity of actors come to develop a category, under which they can group together, albeit with different meanings attached, in a trading zone. This category becomes institutionalized due to resonance of meaning in the different fields analyzed. This is when elements of multiple network domains fold together. Once this category is established, economic actors can cognitively attach themselves to this new category. In turn, the new category helps to promote new actors to raise funds and to invest, which bolsters the maintenance of the new market of targeted breast cancer therapies.

Economic Sociology and the Processual Study of Innovation

The larger theoretical framing of the book brought together two strands of theoretical discussions in sociology that view the social world as constituted of relational ties, be it ephemerally social or observational. I have argued that an analytical perspective of relational sociology, which combines elements from White's social theory as developed in *Identity and Control* with those from actor-network theory, provides insights into the processual constitution of new social formations. Accordingly, the book has viewed relations, interpretations, and actions as mutually generative, and, importantly, has translated the theory into an analytical model to formally model and analyze these sociocultural processes.

The book started with the analytical problem of how to study innovation and the emergence of newness from a processual perspective, which is a central question if we want to better understand contemporary economic life. Building on the larger framework, the book combined elements from cultural sociology with those of economic sociology to yield insights on how a market emerges. As a result, it offers a cultural analysis of the emergence of a market.

Overall, the book presents an empirical contribution to the field of relational sociology, an approach that bridges core theoretical and methodological rifts in the discipline. In doing so, the book makes two further contributions to ongoing discussions in sociology.

First, the book contributes to recent developments in economic sociology toward the study of cultural processes when it focuses on the role of stories to explain how markets emerge. It complements a solely structural perspective on the emergence of new social formations by taking sociocognitive and sociomaterial actors and their stories into account.

More specifically, the book asked how actors deal with ambiguities of newness. In answering this question, it has linked discussions on market emergence with those of collaborative meaning-making. The argument has been that due to collaboratively established interpretations, actors can reduce ambiguities of newness and mobilize financial resources. How economic actors talk about themselves positions them in a larger relational context. Their stories and actions are thus *inter*dependent of others, while they may not agree on interpretation or share understandings. They are cognitively and narratively connected. We have seen that actors construct categories through collaborative, narrative processes to help them figure out what this is about, what they are about, who their competitors are, and how they can be evaluated.

At the same time, such stories tell about the "socially structured imaginaries about worth" (Fourcade 2011: 1769), in particular what *targeted therapies* means in that nascent market. The analyses found multiple and heterogeneous criteria of worthiness, in which patients, scientific advancements, molecules, and mechanisms are evaluated and used as testing agencies. Actors in the nascent market tell stories about the worth of *targeted therapies* when they relate to the number of patients that will be cured. They allude to a "potential wonder drug" or a "blockbuster" and thus connect a molecule or a biochemical mechanism with all those hundreds of thousands of women who may be cured. Other examples can also be found in market research reports, which point to "growing patient's market" and its "profitability." Actors in the nascent market relate the worth of *targeted therapies* to the scientific achievements or medical improvements. At the same time when suggesting economic successes, researchers may also talk about the scientific successes their research strategy exhibits. This is exemplified in press releases, in which a researcher speaks toward investors but at the same time toward the scientific community (and their competitors) about successes and failures. Actors in the nascent market also relate the worth of *targeted therapies* to the newness of the product. This might refer to new as in "first mover advantage," a particular quality ascription, and thus speak to the structuring element of time in this process. Or they indeed speak about new and novel in the sense of something that does not fit into the existing categories of the particular context, the uncategorized. Prime examples are when journalists and business analysts try to make sense of a biochemical mechanism using the language of expectations.

The analysis thus combines a focus on market-making and meaning-making when it highlights how categorization of innovations in ambiguous settings comes about. This book traces processes of categorization, beginning *before* a category exists around which economic actors can cognitively rally and pursue their economic interests accordingly. In tracing categorization processes, their contestations, discursive struggles, and eventual adoption, the analysis found three phases: A first phase began in the late 1980s until first product approval. A second phase began after the first product approval, when *targeted therapies* as a label was discursively contested. The study showed discursive struggles among firms and intermediaries if the new approach and its small molecules are part of a prior category or if indeed they form a separate product category, onto which other biotech and economic actors can legitimately attach themselves. A third phase began after *targeted therapies* was taken for granted as a category according to which companies could sort themselves and their research. With the new category a new market of *targeted therapies* was able to form, which significantly altered the economic value creation models of the pharmaceutical industry. In the emergence of the new category *targeted therapies*, the analysis also showed resistance and refusal in particularly from intermediating industry analysts to use a new label and recognize the new category and thus confirms that "categories that seek to alter radically a profession's logic are likely to encounter stiff resistance, because the new category also alters identities, interests, and statuses" (Jones et al. 2012: 1539). The empirical analyses ended with the year 2010, at which time three targeted breast cancer therapeutics—*trastuzumab*, *bevacizumab*, and *lapatinib*—had been approved. By showing how a diversity of actors collaboratively, over time, in each context tell stories about the meanings of molecules, research strategies, and other economic actors, this book advances work on the categorizing of innovation. In this sense, this book further contributes to research viewing categorizing as a social process (Durand et al. 2017) with distinct phases (Grodal/Kahl 2017; Slavich et al. 2020) because of shifts in contested meaning-making processes. In addition, it expands the "vocabulary structure" approach (Loewenstein et al. 2012), which views words as carriers and generators of meaning, to methodologically encompass how large systems of meanings, i.e., networks of co-occurring terms, come to form a new category that gets institutionalized. In showing category emergence, *Making Sense* contributes to ongoing discussions of cultural sociologists and organizational scholars alike on the basic social processes of valuation, evaluation, and the ascription of worthiness when sorting, classifying, and categorizing occurs.

A second contribution to ongoing sociological discussions is methodological. The book offers a relational and processual perspective to study dynamics of market emergence over time. To do so, the book builds on works that "measure meaning structures" (Mohr 1998; see also Edelmann/Mohr 2018; Mohr et al. 2020; Rawlings/Childress 2021), using a combination of methods. The analyses combine in-depth qualitative text analyses with novel, computational methods of text analyses, i.e., topic modeling and semantic network analysis, to enable a sociological inquiry into meaning-making over time. The mixed-methods research design limits the shortcomings of either only qualitative or only formal inquiries when using texts as data. Certainly, formal analyses reduce complexities without capturing all cultural nuances. The large-scale text analyses employed here are sensitive to contextual semantic meaning-making although they too reify reality. While they allow for detecting patterns over long periods of time across many sources, the qualitative inquiries focused on much shorter periods with a much more limited number of sources. Yet the qualitative inquiries were able to capture fine elements of interpretation that were not detectable with the large-scale analyses. Together, qualitative and formal analyses of textual data show elements and mechanisms of the interpretative and evaluative construction of a market. Such an approach thus promotes a methodological strategy for an empirical analysis of how narratives help to construct a market and, more generally, how cultural elements constitutes social networks. Moreover, such a design that combines interpretive, human skills with computational, pattern recognition ability permits one to overcome the rift between qualitative and quantitative approaches in sociological analyses, to produce more rigorous insights. In this quest, this book joins current research efforts (e.g., Bail 2014, 2016b; Karell/Freedman 2019; Nelson 2020, 2021).

Future Research

This book has considered a high-risk, high-gain market of potentially life-prolonging innovations as a showcase. But the model of market emergence, with its elements of stories of the future, expectations driving the construction of the market, and institutionalization of a category by network folding, can also be applied to other nascent markets.

For example, further research could apply the model to emerging markets, in which it is unclear what the product is about and who will participate. This pertains particularly to markets of scientific, technological, and artistic innovations. Digital, nontangible innovations, to which promises of an improved future may be narratively attached, are fitting candidates for such analyses. For example, the technological innovation of payment transactions using smartphones has been heralded as a game changer for banks, marketing, platform companies, and customers (Mützel 2021). In the emerging market that may constitute the backdrop for "the metaverse," stories of the future are currently driving technological developments, stirring expectations and shaping investments from industries including but not at all limited to education, urban planning, fashion, and advertising (Herrman/Browing 2021). Similarly, research focused on diverse sets of actors involved, including the domains of science, technology, industry research, and marketing, could trace innovations such as driverless cars, biodegradable plastics, electronic currencies, or mRNA vaccines from the perspective of stories that help the involved actors to cope with uncertainty and that, in turn, shape their markets. Further research could also focus on other, nontangible innovations using the book's analytic model, including the emergence of new genres, new styles, or new professions that stories of the future help to shape and that a diversity of actors narratively compete about.

Moreover, the methodological approach this book offered, using large-scale, computational text analyses in combination with qualitative inquiries for when meanings are unclear, is applicable to other processes of emergence in order to understand how the meaning of a category came about. Sociological analyses can profit from using the combination of such tools because they allow the gaining of insights into processes and mechanisms.

Into the Future: Cancer Moonshots

Beyond the field of sociology, the book has also made a substantive contribution to studying developments in a scientific field of innovations: it told the recent history of developments in breast cancer therapeutics research. It gave a polyvocal perspective of how topics of cancer research have changed over time, one that emphasized that the interactions of a multitude of actors were essential to bringing about this change. The empirical analysis found three phases: a first phase began in the late 1980s until first product approval; a second phase began after approval, when targeted therapies was categorized after contestation, and closed by the mid-2000s; a third phase began after *targeted therapies* was taken for granted as a category according to which companies could sort themselves and their research. The empirical analyses ended with the year 2010, at which time three targeted breast cancer therapeutics—*trastuzumab*, *bevacizumab*, and *lapatinib*—had been approved. Analyzing categorization processes of innovation thus also helped to periodize developments.

In 2012, an international consortium of cancer researchers significantly contributed to a better understanding of the molecular profile of breast cancer in particular (Cancer Genome Atlas Network 2012): they established that there are four genetically distinct types of breast cancer: HER2+, hormone receptor+, BRCA1/BRCA2, and triple-negative breast cancer. These types differ from the categories used in clinical research until then (Maxmen 2012b). One type identified, so-called triple-negative breast cancers (TNBC), which are neither estrogen receptor positive, progesterone receptor positive, nor HER2 positive, was found to resemble ovarian cancer and a type of lung cancer (Kolata 2012). About 10–20% of invasive breast cancers are TNBC and typically treated with chemotherapies.[5] Moreover, some BRCA1 and BRCA2 mutations have been found to correlate with TNBC (e.g., Foulkes et al. 2010; Lord/Ashworth 2016). These insights into the genetic makeup provide for new expectations: they could potentially translate into game-changing developments in term of research strategies that would transpose the market; they could also help the hundreds of thousands of breast cancer patients for whom noninvasive treatments do not exist so far. Currently, new categorization processes can be detected, which are shaped by novel genetic insights and concerted efforts for scientific progress (Nik-Zainal et al. 2016; Tan et al. 2020). In 2019, and thus twenty years after Herceptin was first approved as a target therapy, the breast cancer market totaled US$ 20.2 billion; it is expected to grow 9% annually to US$ 47.7 billion until 2029 (Wilcock/Webster 2021).

Since so much of this book is about stories of the future, in which uncertainties are suspended and past and present insights converse with the future, it is only fitting to catch up with recent developments in breast cancer therapy research in concluding the book. Indeed, for the four types of targeted therapies—therapeutics for HER2 + breast cancers, for hormone receptor+

breast cancers, for BRCA gene mutations, and for triple-negative breast cancers—a total of eighteen targeted therapeutics have been approved by regulatory agencies until the end of 2021. Seven of them received approval in 2020 and 2021, which suggests a new phase for breast cancer patients. For HER+ breast cancer, two mAbs (*pertuzumab*[6] and *margetuximab*[7]), two mAbs with chemotherapeutic agents (*ado-trastuzumab emtansine*[8] and *fam-trastuzumab deruxtecan*[9]), and two kinase inhibitors (*neratinib*[10] and *tucatinib*[11]) have been approved. Progress has also been made in developing targeted treatments for hormone receptor positive breast cancers. *Palbociclib*,[12] *ribociclib*,[13] and *abemaciclib*[14] block proteins in the cells (so-called CDK4/6 inhibitors), which helps stop cancer cells from dividing. Also, treatments blocking other proteins in hormone receptor positive breast cancers have been developed: *everolismus*[15] blocks mTOR; *alpelisib*[16] is a PI3K inhibitor. For women with BRCA mutations, PARP inhibitors have been developed and eventually also approved: *olaparib*[17] and *talazoparib*.[18] For triple-negative breast cancer, one immunotherapeutic, monoclonal treatment, *pembrolizumab*,[19] and one treatment with a novel target, *sacituzumab govitecan*,[20] has been approved by the end of 2021.[21]

This brings the total of targeted therapies for the treatment of breast cancer to eighteen at the end of 2021, still a small number, especially in light of the vast number of biochemical compounds that are in preclinical or clinical trials during any given year in the period under study (e.g., Lawrence 2007b). Indeed, the failure rate in anticancer drug development in general continues to be higher than for other diseases. Only about 5% of potential anticancer compounds move through phase III clinical trials and are eventually licensed. This compares to, for example, a 20% success rate in cardiovascular diseases (Ocana et al. 2011). Throughout the years, research has identified a plurality of reasons for this: a general lack of understanding of the complexity and heterogeneity of cancer, including the biology of metastasis vis-à-vis early tumorigenesis (Eckhardt et al. 2012) and the mechanism of action of a given drug, bringing chance into the development process (Gladwell 2010); a focus on targets and therapies that have already shown success, what one industry analyst has compared to a "march of the lemmings" (Booth 2012); a general lack of R&D productivity (Pammolli et al. 2011); and a mismatch of clinical trial protocols and molecules tested (e.g., Buonansegna et al. 2014; Keating/Cambrosio 2012).

While insights into the genetic makeup are advancing, other hurdles in breast cancer treatment research persist. Researchers in particular point to the need to share more information on biomedical research, clinical trials, and patient data; one idea is to turn cancer research into a more data-sharing endeavor, in which findings can be linked and analyzed computationally (e.g., Yao et al. 2010). While open-science approaches (Maxmen 2012a), collaboration (Lengauer et al. 2005), and new financial models (Fernandez et al. 2012) have been discussed for several years, as have earlier expectant turning points on "the war on cancer" (Haber et al. 2011), in late 2016 the US government launched a new initiative, a "cancer moonshot" (Lowy/Collins 2016), to help push science from the "cusp of breakthrough" to an actual breakthrough. In concerted efforts, knowledge on molecules, pathway, targets, therapies, patients' genetics, and their treatment responses are to be shared between scientists for the development of better treatments in a Human Tumor Atlas Network (Rozenblatt-Rosen et al. 2020). New databases, such as the Cancer Research Data Commons (CRDC), with standardized and shareable information as well as new efforts in funding, are to foster scientific breakthroughs. While progress has been made, as the recent produce approvals attest (Sharpless/Singer 2021), "efforts must be scaled up, redoubled, and accelerated," to lessen the burden of cancer worldwide (Agus et al. 2021: 165).

In these efforts, "no simple, universal, or definitive cure is in sight—and is never likely to be—the past is constantly conversing with the future. Old observations crystallize into new theories" (Mukherjee 2010: 466). New findings will raise new possibilities and expectations, reinterpret past insights, and bring about new futures—and new markets from stories.

Appendix A:
Research Design and Data

The analytical focus of the book is on multiple network domains, a heterogeneous set of actors, and their interplay: biotech companies, which at some point in time claim that they are working on breast cancer therapeutics; financial and business analysts, who evaluate the biotechs; journalists, who write about the biotechs and their research; the scientific oncological community; and molecules, compounds, and biochemical mechanisms. Notably absent from this list are political decision-making bodies, regulatory agencies, as well as patients and their organizations. These have been studied elsewhere with great expertise (e.g., Charon 2006; Keating/Cambrosio 2012; Klawiter 2008; Kolker 2004; Rabeharisoa/Callon 2002). They enter into the stories, though they are not focus per se of this book's investigation.

The data the book uses are texts. They cover the period between 1989 and 2010. My motivation to use "text as data" (see also Grimmer et al. 2022; Grimmer/Stewart 2013) is driven by several considerations. Texts written by the actors of interest allow analyzing what was happening as it was happening, as opposed to, for instance, *ex post* interviews, which would necessarily involve a sampling of sorts and add issues of recollection. Since this study is less interested in individual actors and their attributes than it is in events and actors in their interactions and dynamics over time, primary texts are the type of data of choice to get at the description and analysis of processes. They hold the public stories of facts and expectations this study is interested in. Moreover, from a methodological point of view, texts can be analyzed qualitatively and with more formal methods.

To capture the different network domains and actors and to get at a plurality of perspectives, several archival data sources had to be tapped. In addition to following developments in the field of biotechnology and breast cancer therapies by regularly reading magazines such as Nature's *Bioentrepreneur* and *Biotechnology*, and having informal conversations with oncology experts and members of the biotech community, I constructed six original data sets from archival databases for further systematic analysis.

In a quest to understand the early years of innovative breast cancer therapeutics, I constructed a data set that combines companies' press releases, journalists' reports, and industry analyses.[1] I used the commercially available full-text database LexisNexis Academic Universe to collect all articles that included the terms "biotech* AND cancer*" beginning in 1989.[2] Such a search allows for an inclusive analysis of what was going on in the field and fits the theoretical insight that we cannot ascribe *a priori* which actor or which molecule are important when studying the developments of innovation processes. The search originally encompassed the years 1989 through 2000. After a first round of analysis and because of the events happening, I reduced the time frame to 1989 through 1998. This data set consists of n=845 articles.

A second data set contains a series of commercially available reports from the market research company Business Insights from 1999 to 2011, specifically on the cancer market as well as on relevant fields of biomedical and oncological research developments. These reports evaluate the current state of the market, provide forecasts, discuss important companies, and highlight market opportunities. In turn, they are used as data sources in other publications, e.g., *Nature Reviews Drug Discovery*. Typically, these reports are book-length treatments of two hundred or more pages. Twenty-three such reports form the basis for this data set.

In order to understand scientific developments of the field of oncology, I searched the commercially available database Web of Science, Science Citation index, using the search string "(breast* OR mamma*) AND (cancer* OR tumor* OR neoplas*) AND therap*" within the Web of Science subject category "oncology." I collected every publication entry, including abstracts, for the period between 1989 and 2010, amounting to n=31,070 documents in total, from a total of 221 journals. This constitutes a third data set.

To capture the other network domains, I again used the commercially available, full-text database LexisNexis Academic Universe and collected every article that used the terms "biotech* AND cancer*," now for the period between 1989 and 2010, in three types of sources separately. My specific source selection is based on relevance for the field, in an attempt to capture different geographic nuances of the field, and on availability over the period 1989–2010.[3] I group

Table Appendix.1. Overview of Data Sets

Type of Data	Sources	Search String	Digital Archive	Date Range	Total # of Documents	Total # of Words
Press statements, newspaper articles, reports from industry analysts	wire news; international, US, UK newspapers; biotech news	biotech* AND cancer*	LexisNexis	1989–1998	845	
Market research reports	Business Insights			1999–2010	23	
Scientific discussions, abstracts	subject category "oncology," 221 different journals	(breast* OR mamma*) AND (cancer* OR tumor* OR neoplas*) AND therap*	Web of Science, Science Citation Index Expanded	1989–2010	31,070	7,631,888
Reports from industry analysts	*Pharma Marketletter, Biotech Business Week*	biotech* AND cancer*	LexisNexis	1992–2010	93,887	44,887,578
Wires, press statements	PR Newswire (Europe, Asia, US), Business Wire	biotech* AND cancer*	LexisNexis	1989–2010	12,672	11,279,360
Newspaper articles	*New York Times, Financial Times*	biotech* AND cancer*	LexisNexis	1989–2010	1,073	961,966

the sources according to their own classification: reports by industry analysts, wire reports and press statements, and newspaper articles. Industry analysts are biotech news providers that translate the claims of companies for a heterogeneous audience of investors and industry participants. The data set of industry reports between 1989 and 2010 comprises n=93,887 articles. Wire reports and press statements of companies worldwide are publications consisting of news about a given company and bring new developments into the public domain. The data set of wires and press statements consists of n=12,672 articles. Newspaper articles of the *Financial Times* and the *New York Times* are written for investors as well as a general audience. The data set of newspaper articles contains n=1,073 articles. Whereas the data set of Web of Science publication entries captures scientific findings, the latter three data sets evaluate and interpret research results or new company developments. Table Appendix.1 presents an overview of the different data sets, with sources selected, search strings, and further specifics.

The data set construction thus includes four very large data sets and two smaller data sets. Two methodological choices follow:

The two smaller data sets permit a close qualitative, content analytic reading of how competitors listen, observe, and react to each other and how business and media analysts interpret and evaluate. Such an interpretive reading of texts and a tracing of their interrelatedness will allow gaining insights into the proposed collaborative meaning-making processes (Chapter 3 and Chapter 4).

The larger data sets pick up the current challenge in the social sciences to deal with complete sets of archival textual data. In the spirit of coping with the "coming crisis of empirical sociology" (Savage/Burrows 2007) and contributing to the "watershed moment in the social sciences" (McFarland et al. 2016), in which data and methods are being dramatically expanded in what is often referenced as *computational social science* (e.g., Edelmann et al. 2020; Evans/Aceves 2016; Lazer et al. 2020; Lazer et al. 2009; Watts 2013), the analyses presented in Chapter 5 use novel computational approaches to text analysis. Using "old big data" (Bearman 2015) from digitized archives, the methods of topic modeling and semantic network analysis will be used to discover patterns in the texts. Both approaches are based on computational linguistics. Whereas the first makes use of machine learning techniques and unsupervised modeling, the second is rooted in the social network analytic tradition. Both detect similarity and structural association of textual details that, in turn, can provide for meaningful interpretation by the analyst. Moreover, they offer systematic insights into dynamics and trajectories of large-scale processes *without a priori* ascriptions of what is relevant in the field. While topic modeling is a useful method to discover categories from a corpus of texts over large periods of time, the semantic network analysis employed here can show drifts and shifts in the structure of terms making up the categories over time using network analytic algorithms. Chapter 5 includes nontechnical discussions of the two methods used. Technical details are given in Appendix B.

The study thus combines three methods of text analysis to capture meaning-making processes: close content analytic readings of observing and reacting competitors and their analysts (Chapter 3) and of market analysts (Chapter 4); large-scale text analyses of developments using topic modeling as well as semantic network analyses in the fields of oncology, the biotech business, as well as their biotech industry and finance analysts from a macroscopic perspective (Chapter 5).

These methods translate foundational theoretical ideas: through their own texts, such as press statements, biotech companies position themselves, while they are also being narratively positioned by financial and business analysts' interpretation and evaluation. Similarly, the scientific community positions molecules, compounds, and biochemical mechanisms, and clarifies their worth for research and business. This can be captured by qualitative analysis. Such qualitative analysis, which has the relational at its core, is then combined with novel methods based on computational linguistics, machine learning, and network analysis to detect patterns rooted in relations of terms over time.

In designing the study, I made several critical choices. The first one was to focus on the entire field for a long period of time. This choice was motivated by the interest to analyze the process of emergence over several years. Any sampling of texts would hinder delineating dynamics and trajectories. A second consequential choice was to be most inclusive with search strings used in data collection. Rather than already focusing on, e.g., a particular biochemical mechanism, the search I conducted included every article that had anything to do with "biotech and cancer." This yielded a wide array of texts, in which the role of biotech and cancer could be central but could also be marginal. The third choice was to obtain data from different types of sources. This was motivated by theoretical insights that multiple network domains are involved in forming a market. While an interest in trajectories drove one part of data collection and analyses, a fourth critical choice was to combine large-scale analyses with content analytic readings of texts. This allows capturing closely detailed interactions, while the macroscopic view allows seeing the larger picture.

Appendix B:
Technical Details on Formal Analyses

Appendix B provides further technical details on the formal analyses in Chapter 5. I point to sources that provide further mathematical details.

Topic Modeling

Topic model analyses encompassed four different textual corpora: scientific discussions, industry analysts' reports, wire reports, and newspaper articles. The data were gathered using digitized archival sources (LexisNexis and Web of Science). For an overview of how the corpora were selected, and how many documents and words they contain, see Table Appendix.1.

Once collected, the data set construction continued with processing the texts. In addition to automatic procedures, I worked with text editors and corpus analysis toolkits to look at the data. To prepare each corpus for analysis, linguistic processing was necessary. For each corpus, this entailed removing stop words, i.e., *the, and, because,* all numbers, punctuation, and white spaces, the most common words in each corpus, i.e., *breast, cancer, patients* in case of the scientific discussions, words rarely used, words with less than two letters, and documents with fewer than five words. This also included harmonizing different spellings, e.g., *F.D.A.* and *FDA* became *fda*; all letters were transformed to lower case. N-gram analyses suggested to pull compounded terms together, e.g., *unitedstates, forwardlooking, britishbiotech*. Other common practices of natural language processing (NLP) such as lemmatization (identifying the base form of a word) proved difficult for the purposes of analysis. For example, *pharmaceuticals* can mean to denote products or companies and thus could not be simply shortened to *pharma*. The result of the cleaning was an article-by-terms-matrix.

The corpora were then analyzed using a commonly used, simple topic model algorithm: symmetric Latent Dirichlet Allocation (LDA) with Gibbs sampling (Blei 2012; Blei/Lafferty 2009; Blei et al. 2003). For formal modeling I used the publicly available Stanford Topic Modeling toolkit developed by the Stanford Natural Language Processing Group.[1] In addition to the documentation online, Graham et al. (2016) provide how-to-instructions. Ramage et al. (2009), McFarland et al. (2013), and Kaplan and Vakili (2014) provide further technical details.

Before running the analysis, the analyst must decide on the number of topics. The decision is at once driven by a quest for interpretability as well as how close or far the topic model algorithm should adjust the lens to look at the data. For all corpora, I ran multiple models with different numbers of topics: 30, 50, 100, 150, 200, 250, 300 topics. For three large-scale corpora I found the 200-topic solution best for interpretation. For the relatively smaller corpus of newspaper articles, I opted for a 50-topics solution.

Another decision the analyst has to make prior to analysis is that of the size of the parameters α and β, which are set above 0 and typically not much larger than 1. The α-parameter represents document-topic density. A low α-parameter means documents contain fewer topics, i.e., a focus on fewer but dominant topics; a high α means that documents contain many topics. The β-parameter in turn represents the topic-word density. A low β-parameter means that topics consist of few words of the corpus, i.e., those that are dominant across the corpus; a high β means that topics consist of most words of the corpus. In the conducted analyses I set both parameters to a low number, i.e., 0.1 (see Wallach et al. 2009 for a discussion on priors).

The topic models were estimated using Gibbs sampling using a Markov Chain Monte Carlo (MCMC) technique (Steyvers/Griffiths 2007), which I ran for 2,000 (1,000 in the newspaper corpus respectively) iterations. I then labeled the topics. For the scientific corpus, I consulted cancer researchers to verify my labels and to clarify some terms. For all other corpora, I consulted with research assistants on the labels.

Semantic Network Analyses

For the semantic network analyses, I used the corpus of scientific discussions gathered from Web of Science, as used in the topic modeling described above and as shown in Table Appendix.1. I used CorText, a Python-run platform for the analysis of textual heterogeneous networks, developed at the IFRIS in Paris (http://cortext.net/). In its implementation it combines tools of NLP with network analytics. The CorText platform offers background documentation and tutorials (https://docs.cortext.net/). For further computational details see especially Rule et al. (2015).

Several analyses in Chapter 5 are based on frequently occurring noun terms, including multiterm phrases such as "growth factor," "tumor cells," or "monoclonal antibody," which were extracted from the abstracts using NLP techniques that are part of the CorText platform. After reading in the data, by which the data gets parsed and transformed into a database, TreeTagger (Schmid 1995) tagged each word in the corpus according to its grammatical category. The corpus was then chunked to identify noun phrases with a maximum length of three words. A stemmer reduced and unified varieties of terms to the same term so that, e.g., both "antibodies" and "antibody" became "antibody." Also, semantically coherent multiterms, such as "cancer therapy," "therapy of cancer," and "therapy against cancer" were pulled together under the same heading, in this case "cancer therapy." CorText then provided an index of terms and their frequency in the corpus. From this list, I selected the 1,000 most frequent multiterms for further analyses.

The illustrative analysis of terms and research organizations (Figure 5.14) uses those extracted multiterms and relates them to a category from the Web of Science input on research organizations. This analysis is based on an affiliation matrix of terms and research organizations.

All other analyses are based on co-occurrence patterns between *terms* over time (extracted multiterms from the abstracts or keywords as provided by authors). For those terms-by-terms analyses, a first step is to identify the co-occurrence of term i with term j in each abstract and based on the list of terms extracted and prepared earlier. A proximity score then computes the relatedness of each pair of terms, yielding a semantic network that is weighted according to contextual measures, in which terms are the nodes and edges denote their similarity. CorText filters the edges to retain a connected network, which in turn means that loosely connected terms are dropped. The Louvain community detection algorithm (Blondel et al. 2008) then identifies cohesive subgraphs, which I interpret as categories of scientific discussions.

The networks are mapped using a classic force-directed layout. The nodes are colored according to the cluster to which they belong. The circles over each cluster are scaled in size according to the number of abstracts that have been assigned to this category. The mapping of each cluster allows for insights into the terms making up each cluster and also, more importantly, how each cluster is structured and how clusters are connected to each other. The size of each node speaks to the frequency of the term in the corpus.

Based on all extracted words, a global map (Figure 5.15) shows the semantic network structure of oncological discussions on breast cancer therapy between 1990 and 2010.

Because the detectable categories are rather broad, I also extracted keywords that are attached to abstracts in Web of Science to get at shifts in treatment categories, global and broken down by different periods over time. The global map of extracted keywords (Figure 5.16) is based on the top 100 keywords and their relations to each other for purposes of better readability. I reduce the number of keywords to the top 50 in the analyses of five different time periods, which I set according to the previous insights into the developments of the field (Figure 5.17).

In a third part, I conducted a lexical extraction of noun phrase multiterms with maximum length of three with the pivotal term "target" at the sentence level. In each extracted multiterm, then, "target" appears. As described before, semantically coherent multiterms were pulled together; a stemmer reduced variances. I then used a periodization tool implemented in CorText that compares average similarities between the top 1,000 terms used in abstracts per year. Each cell in the transition matrix (Figure 5.18) compares the term vectors for two years of abstracts, showing their dissimilarity on a scale from 0 (most similar) to 1 (most dissimilar).

The global map (Figure 5.19) using the top 100 terms and the periodized maps (Figure 5.20) using the top 50 terms on the discussion of "target" were constructed according to the descriptions above. Using the same five periods and on the basis of a computation of distances between pairs of temporally consecutive clusters, a tube (Chavalarias et al. 2011) or Sankey diagram (Figure 5.21) can be constructed that identifies how the categories evolve over the time periods. The size of each category bar again marks the number of abstracts that are assigned to each cluster based on their terms.

Notes

Introduction

1. To be sure, for hundreds of thousands of breast cancer patients the so-far approved, noninvasive therapeutics remain ineffective, or they develop resistances to the treatments. The search for treatments is ongoing.
2. This is a dynamic much like the interaction preludes of Leifer (1988).
3. See Appendix A for more on data sources used.
4. The notion of "relational sociology" may appear strangely redundant. After all, the discipline of sociology is rooted in the study of social relations. In his work on the division of labor, Durkheim pointed out that societies differentiate themselves according to the structure of their social relations. Marx famously pointed out that a lack of social relations between French small-holding peasants in the nineteenth century explained why they did not constitute a class. Simmel saw that people and the groups they belong to were not only interacting but indeed formally interdependent as "intersections of social circles" (see Mützel/Kressin 2021 on Simmelian roots for a relational sociology). The statistical revolution in the social sciences of the post–World War II decades established variable-centered approaches that study relations between variables (Abbott 1988) rather than relations between actors. So, the issue is what types of relations do we mean when we think of sociology as the study of social relations. At the same time, "relational sociology" is a trending term in the international market of ideas. It has come to connote different approaches. Some refer to a structuralist, network analytic approach in the study of art worlds as relational sociology (Crossley 2010, 2015). Some outline a "relational theory of society" (Donati 2010, 2015). Others focus on Nobert Elias's transactional approach (Dépelteau 2015). Still others propose to build relational sociology on Niklas Luhmann's concept of communication (Fuhse 2015a, 2015b, 2021). This book builds on existing discussions (e.g., Abbott 2007b; Emirbayer 1997; Fuhse/Mützel 2010) while also offering an empirical contribution.
5. This section builds on ideas of Mützel (2009).
6. These ideas have been productively expanded for a range of fields, e.g., Emirbayer and Johnson (2008), Mutch et al. (2006) on organizational analysis, Moody and

White (2003) on structural cohesion, and Lamont and Molnar (2002) and Wimmer (2008) on the study of boundaries.

7. White (1992) uses the terms "stories" and "narratives" interchangeably.

8. A momentary successful translation results in blackboxing. When a whole entanglement is mobilized in the name of one problem defined by one actor, its heterogeneity and distributed character is blackboxed, so that it seems that one bounded actor carries out the action. Indeed, ANT understands itself as a research program oriented toward opening up blackboxes that hide the hybrid constitution of the social world.

9. Technically, such maps present multiple affiliation ties in the same space. The algorithm estimates measures of centrality, betweenness, and equivalences and thus positions individual nodes accordingly. These studies thus bring two mutually associated sets of elements together in the same visual representation of networked relations, in order to show processes of collective collaboration. This ANT version of network analysis was originally developed using the software ReseauLu (Cambrosio et al. 2013; Cambrosio et al. 2006); currently CorText presents a powerful option (cortext.net; see, e.g., Bourret et al. 2014; Chavalarias/Cointet 2013; Rule et al. 2015; Venturini et al. 2014).

Chapter 1

1. White uses the idea of niche as it is applied in organizational ecology and links it to the network theoretical concept of social position. Market niches emerge in a competitive process and define the identity of a company. The process of finding one's niche then allows the market structure to be reproduced.

2. White's model maintains that consumer preferences are not fully predictable. As aggregate decisions reflecting on the quality of products, consumer decisions contribute to the reproduction of market structures in the next observation interval. Consumers are able to distinguish between individual producers and the quality of their products, whereas producers cannot observe individual consumer preferences *ex ante*. It is only after products are on the market and all the related decisions have been made that producers can judge by the customers' reactions whether they followed a successful strategy or not.

3. The fundamental building block of such structural comparison has its origins in the concept of *structural equivalence*. It analyzes the patterns of relationships that define the position and role structure of one actor relative to all other actors in the network. In the original version (Lorraine/White 1971), two actors are considered to be structurally equivalent if they have identical connections to and from other members of the network. They do not have to be directly connected to one another and they do not have to belong to the same clique. What counts is the similarity of the relational patterns to other actors in the network. Blockmodeling is the algorithmic implementation of this concept (Breiger 1976; White et al. 1976).

4. For empirical analysis, White suggests specific computations, the $W(y)$-model, to capture the underlying, cost, volume, and quality structure for market profiles of existing markets.

5. To be sure, there have been earlier sociological studies of financial markets that focused on relations and cultures (e.g., Abolafia 1996; Baker 1984). Other seminal studies on financial markets had also shown the role of objects and knowledge (e.g., Knorr Cetina/Bruegger 2002) indicating that these types of markets are "scopic markets," i.e., fully visible on one screen, as opposed to a small group of observers watching each other or direct interactive exchanges (e.g., Knorr Cetina 2006). Financial markets in that sense are fundamentally different than producer markets (e.g., Knorr Cetina/Preda 2005). Though similarly highlighting the role of objects in the social study of finance, the latter perspective does not adhere to the ANT program.

6. Applying this research program, e.g., Çalışkan has studied a global, agricultural market (2010).

7. In discussions with White's market model, Favereau et al. (2002) suggest that markets come from quality conventions. However, it is not one convention of what constitutes quality, as assumed in White's market model, but rather a plurality of quality conventions that come to structure and form markets. These conventions of what constitutes quality are established among market participants based on cultural schemata.

8. Boltanski, Thévenot, and their collaborators have identified eight orders of worth or "worlds," on which evaluations in critical situations are based (the world of the market, industrial world, civic world, the world of fame, domestic world, the world of inspiration, the green world, the project-oriented world) in the regime of justification. These orders of worth cut across social domains, such as economy, family, religion, and science. The world of the market, in which the main quality is to seize opportunities and to be competitive, can not only be found in the economy but also in other social realms, such as in science or in the arts.

9. Firms are solutions to the coordination of multiple orders of worth (Thévenot 2001).

10. But not all texts are narratives. And not all narratives analyzed are analyzed as narratives but perhaps only elements thereof. Also, as one seminal book on qualitative data analysis methods points out "most of the analysis is done with words. They can be assembled, subclustered, or broken into segments. They can be reorganized to permit the researcher to compare, contrast, analyze, and construct pattern out of them" (Miles et al. 2013: 9).

11. According to MacKenzie's discussion on the notion of performativity, this would encompass both "effective performativity" as well as "Barnesian" performativity (2006: 17).

12. To be sure, there are varieties of conceptions of "discourse," of which not all work with Foucault (Alvesson/Kärreman 2000; Angermuller et al. 2014). See also Maesse (2013), Tellmann (2003), and Wetzel (2013) for discourse theoretic approaches to the study of the economy.

13. Czarniawska's discussion introduces a "narrative institutionalism" (1997: 193–194) that extends DiMaggio and Powell's "new institutionalism" (1991) by introducing narrative as the key form of cognition, as opposed to "classifications, routines, scripts, and schema."

14. Since 1995, US companies need to flag "forward-looking statements" in their press releases to clarify that these statements are only expectations and that uncertainties apply. See Rajan 2006 and Tutton 2011 for insightful discussions of the role of forward-looking statements. See also Chapter 5.

15. Moreover, Gibson has used the term "foretalk" (Gibson 2011a, 2011b, 2012) to illustrate talk-in-interaction and negotiations involved in producing stories about the future. See also Reichmann (2013) on how economic forecasters produce foretalk.

16. Such stories are not to be confused with a call for more fictional, lyrical elements and techniques in writing sociological texts (as in, e.g., Abbott 2007a; Phillips 1995).

17. Typical language uses hedging, i.e., "expected," "anticipated," "intend," "possibly" (McLaren-Hankin 2008).

18. No consensus of the use of the terms "uncertainty" and "ambiguity" seems to exist in the literature (see, e.g., the discussion in Beckert 2016, endnote 9: 289–291). I use "uncertainty" to capture structural aspects, such as social interactions and other actors, and "ambiguity" to capture aspects of meaning and interpretation. In their seminal work, Lane and Maxfield (2005) refer to the first type of uncertainty as "ontological uncertainty": entities and relations are not known. What I refer to as ambiguity they would refer to as "semantic uncertainty." While uncertainty about relations and actors may lead to flexibility on whom to connect to, ambiguity may lead to potentially multiple interpretations. White uses "ambage" and "ambiguity," referring to social and cultural uncertainty, respectively, as dual concepts (1992, 2008).

Chapter 2

1. The focus of therapies necessarily entails that I will not systematically focus on issues of prevention, diagnostics, national differences in health policies, clinical trials, or patient activism. Several works delineate the history of cancer developments between patients' experiences and diagnostic and treatment developments. Some offer accounts from medical insiders (e.g., DeVita 2015; Marks 2014), others focus on patients' experiences (e.g., Kushner 1975; Lerner 2001), and still others provide histories of particular cancers (e.g., Löwy 2011; Olson 2002; Timmermann 2014). Cantor (2007, 2008) and Pickstone (2007) provide historical overviews of treatments and describe contestations and institutional path-dependencies.

2. Cancer statistics rely on local registries for data. Data quality is unevenly distributed across the world. Not all data for all regions are available. The International Agency for Research on Cancer (IARC), as part of the World Health Organization (WHO), is the global agency to collect and communicate cancer statistics.

3. This overview is based on the historical developments as outlined in the works of Lacroix (2011), Mukherjee (2010), and Wagener (2009). Among the three authors, who do not cite to each other, there are minor divergences as to the roles of particular scientists, discussions that I leave entirely aside for this broad overview.

4. Galen also suggested that excess of "black bile" causes melancholy.

5. This was one answer to Nixon's "war on cancer," a US bill signed in 1971 that led to an increase in funding for cancer research.

6. Approval from the Food and Drug Administration (FDA) is a crucial obligatory passage point. European drug regulatory authorities (European Agency for the Evaluation of Medicinal Products, EMEA, founded in 1993; renamed European Medicines Agency, EMA, in 2004) use slightly different procedures to determine an acceptable point for drug approval (see, e.g., Downing et al. 2012; Tafuri et al. 2014; Trotta et al. 2011 for comparison of regulatory approval practices).

7. The scientific literature points to Paul Ehrlich, "the founder of chemotherapy," to have coined the term "magic bullet." "In Ehrlich's language, magic bullets were molecules that bound to parasites or their toxins with the highest possible affinity and displayed very low affinity, if any, for the "haptophoric" groups (binding groups) of the host. The term was born out of his immunological work but was later also used in connection with synthetic molecules that were aimed at specific biological targets (Drews 2004: 799).

8. Fujimura (1996) highlights the practices and processes involved in turning the proto-oncogene theory into a scientific, widely accepted fact, which won the Nobel Prize in 1989.

9. For Genentech's even earlier years producing human growth hormone (hGH), McKelvey (1996), Hughes (2011) and Powell/Sandholtz (2012b: 418–420) provide illustrative organizational histories. On October 4, 1980, Genentech was the biotech company with the first IPO ever: "By day's end, Genentech—without a single product on the horizon—had a valuation of $532 million" (Powell/Sandholtz 2012b: 381).

10. Genentech was in competition with the biotech Biogen, which supported a Harvard team of researchers that worked in British facilities, as well as Novo Industri, a Danish company and leader in the European insulin market (Hall 1987).

11. See Cambrosio and Keating (1995) for an anthropological and historical study of the techniques and the production of monoclonal antibodies until the 1990s. Marks (2015) updates the historical account of "the quiet revolution" of mAbs until 2014.

12. The name HER2/neu stems from the structural similarity to the *h*uman *e*pidermal growth factor *r*eceptor and the rat tumor called *neu*roblastoma, where it was first identified. Several labs independently discovered the same gene (see Mukherjee 2010: 414; Coussens et al. 1985).

13. See also Slamon et al. (1989) for a reiteration of the original findings.

14. Axel Ullrich, for example, co-founded Sugen (Redwood City, CA; 1991; IPO 1994) (Hughes 2006).

15. At that time, interleukins were developed as immune stimulants as well as immunosuppressants, without, however, having a full understanding "which biochemical pathways the interleukins activate" (Smith 1989: 664).

Chapter 3

1. Historians of life science developments have been using oral history techniques to collect information about developments in the field. For example, several pioneers in cancer research deliver their perspective and narrate from a first-person perspective what happened, when, and why (http://bancroft.berkeley.edu/ROHO/projects/biosci/

oh_list.html). But see also a recent version of the Life Sciences Foundation's website that shows a timeline of developments to account for simultaneous developments (http://biotechhistory.org/oral-histories/).

2. This chapter builds on and expands Mützel 2010.

3. To be sure, as this field of research is moving and advancing over time with new insights into biochemical processes and genetic mutations, these heuristics will only hold for a few years of public and scientific discussions before they need to be updated.

4. The search term in the LexisNexis database was "biotech* AND cancer*" in US and European news sources and industry and wire reports. Appendix A gives a detailed overview of the sources used.

5. A research assistant independently checked the coding results. Only in very few items did the coders disagree. Those disagreements could be clarified.

6. Others have found an "organizational embeddedness" of public relations professionals and financial analysts (Grünberg/Pallas 2013).

7. As I follow the developments through time, I use the present tense, as commonly done in ethnographic studies, to capture what was happening *as* it was happening.

8. In the mid-1990s, there are 475 biotech companies in all of Europe, of which 100 are biopharmaceutical companies developing medical agents. In comparison, of the 1,300 US biotechs, 800 work in the biopharmaceutical area (Ernst & Young 1995).

9. Bazell (1998) describes this as a watershed moment for Genentech, which would make or break the company, as phase III trials are costly, difficult to design, and then difficult to carry out. Genentech decided on three separate trials across 150 sites in the United States, Europe, and Australia. It aimed to enroll at least 450 women with newly diagnosed metastatic breast cancer originally for a double-blind test to prove survival, the gold standard of phase III testing (some would receive HER2, some chemo, some a placebo), at least 200 women whose metastasized breast cancer did not respond to chemotherapy, and at least 200 women who had metastatic breast cancer and refused chemotherapy. Because women were not willing to receive a placebo in the double-blind test and treating doctors were skeptical, recruitment was very slow and the entire test was endangered to fail. The FDA changed its clinical trial rules in spring 1996, by which "companies need no longer prove that a cancer drug improves survival or even quality of life. They simply must show that it shrinks tumors, even if only temporarily" (Bazell 1998: 158; see also Keating/Cambrosio 2012, in particular part 3, for changes in clinical trials with molecular treatments). With a new protocol without placebos, an informational campaign for doctors and nurses, the support of the National Breast Cancer Coalition, and eventually much news coverage (Bazell 1998: 161), the numbers slowly rose. Since 1995, Genentech also offered a compassionate-access program ahead of FDA approval for those women who were ineligible for trials (e.g., because they have had too many rounds of chemotherapy) and admitted women on a lottery basis (130). See Lerner (2001, 2002) for the intertwined history of diagnosis, patients, and treatment choices.

10. When stock prices peaked, British Biotech was, for example, ahead of Biogen, which at that time already had both products and profits.

11. In July 1997, partnering Genentech and IDEC receive FDA approval for *rituximab* (Rituxan/MabThera), a monoclonal antibody, treating B-cell non-Hodgkin's lymphoma.

12. At this time, also, several small biotechs in California are working on p53 mechanisms and other pathways inhibitors (Roush 1997).

13. More of these ups and downs surely exist. Another example would be the story of EntreMed (IPO 1996, Rockville, MD), which focuses on developing agents that stop tumor's blood supply (anti-angiogenesis). In 1997, EntreMed developed methods to artificially manufacture the natural hormone angiostatin (in genetically engineered yeast growing in twenty-gallon vats), which prevents angiogenesis. In November 1997, EntreMed reported positive preclinical findings. Entitled "Hope in the Lab," the *New York Times* published a story on the drug developments: "The new drugs, angiostatin and endostatin, work by interfering with the blood supply tumors need. Given together, they make tumors disappear and not return" (Kolata 1998). After that publication in May, stock prices soared, a phenomenon that was later investigated by economists, showing that "enthusiastic public attention induced a permanent rise in share prices, even though no genuinely new information had been presented" (Huberman/Regev 2001: 387).

14. Theratope, it will turn out later, is one of breast cancer therapeutic compounds that failed unusually late, in phase III of clinical trials. Having enrolled more than 1,000 women at more than 120 clinics in North America, Britain, Europe, Australia, and New Zealand, Biomira and by then partner Merck KG (Darmstadt, Germany) learned from preliminary results that "Theratope did not meet the 'predetermined statistical significance' for two end points: survival and disease relapse" (Zehr 2002). As final results came in, analysts respond harshly: "'This really is the final nail in the coffin, the stake through the heart, the final straw . . .' said Duncan Stewart, portfolio manager at Tera Capital Corp. 'If I were CEO of this company, I would shut it down and give the money back to the shareholders'" (Dabrowski 2003). Yet Biomira's CEO does not give up hope and says in a press conference "'Results showed that Theratope appeared to increase the median survival in a subset of patients,' [Alex] McPherson said, adding that this is preliminary information and hasn't yet been fully analyzed. 'So I cannot say that this subset of patients has hit a level of statistical significance at this point,' he said. 'We are hopeful that our further analysis will confirm that Theratope will be a treatment for women in this subset of the metastatic breast cancer patient population'" (*Toronto Star* 2003).

Chapter 4

1. The first year that these reports were systematically available to me is 1999.

2. As indicated earlier, Theratope failed clinical trials at a late stage in 2002.

3. In 2001, the tyrosine kinase inhibitor *imatinib* (Gleevec; Novartis, Basel, Switzerland) received approval in the United States and the European Union. As a small molecule it stops the growth of cancer cells by blocking a particular enzyme. It is used to treat chronic myeloid leukemia (CML). See also Vasella and Slater (2003) for the story of Gleevec.

4. These are Rituxan/MabThera (non-Hodgkin's lymphoma), Herceptin (breast cancer), and Gleevec/Glivec (chronic myeloid leukemia).

5. This field of diagnostic biomarkers presents a complement to the analysis of therapeutics. For instance, Metzler points to "the plethora of hopes and expectations" that are invested in biomarkers (Metzler 2010: 412), Bourret (2005) points to the collaborative work in the clinics using and evaluating biomarkers, Hedgcoe's work (2004; 2003) discusses the risks and expectations of pharmacogenomics more generally, and Bell (2013) discusses the "measured certainty" that biomarkers seemingly offer patients.

6. This is also supported by shifts that other research on the pharmaceutical industry focusing on cancer treatments in general has detected. Styhre's work (2011) identifies a new institutional logic in the pharmaceutical industry that comes to focus on antibodies and large molecules. In oncology, Sosa (2011) could also trace a shift from chemotherapeutics to antibodies and large molecules.

Chapter 5

1. A topic identified endogenously with topic modeling algorithms can be understood as a category. In the remaining discussion of this section, I will use the wording "topic" as it pertains to "topic modeling."

2. The search string was: "(breast* OR mamma*) AND (cancer* OR tumor* OR neoplas*) AND therap*." See Appendix A for further specifics.

3. I used the freely available Stanford Topic Modeling Toolkit (STMT) (nlp.stanford.edu/software/tmt/tmt-0.4/) using LDA with Gibbs sampling (Ramage et al. 2009; see also McFarland et al. 2013). A tutorial is available online and Graham et al. (2016) provide detailed how-to descriptions. In the past few years more options have been developed, including packages in Python and in R. STMT is no longer developed.

4. In addition to validity, robustness is also an issue in topic modeling. I have used perplexity measures and run different estimations, using 30, 50, 100, 150, 200, 250, 300, 350 topics to be modeled. In this case, 200 topics turned out to yield the most robust and interpretable results.

5. I checked the labels with oncology researchers for verification.

6. See Appendix A for further specifics on data sources.

7. See Appendix A for further specifics on data sources.

8. Beginning with 1996, only 12% of all press releases in the corpus analyzed did not contain a forward-looking statement. These were non-US companies reporting, reports on market analysts' views, or the company's press statement referred to personnel announcements only.

9. See Appendix A for further specifics on the data set.

10. See, e.g., the large literature on interlocking directorates and interorganizational networks. Mützel and Breiger (2020) discuss duality beyond persons and groups.

11. CorText is a platform freely accessible online (cortext.net), written in Python. It is an extension of ReseauLu (Cambrosio et al. 2013), and not only maps heterogeneous networks but also allows for analysis of dynamics over time using tools of NLP and network analysis.

12. The global semantic network structure using 500 nodes and 3,748 edges as relations between them resulted in nine clusters: one small cluster on stem cells entered between "quality of life" and "apoptosis," another small cluster on "tumor markers" between "targeted therapies" and "chemotherapy." For purposes of visual readability, I use the seven-cluster solution.

Conclusion

1. In the past decade and with more knowledge on their biochemical mechanism, MMPs have once more received increased attention in cancer therapy research (e.g., Roy et al. 2009; Winer et al. 2018).

2. This practice is also known as "pharmacogenetics," a term derived from "pharmacology" and "genetics."

3. For targeted therapies to exist, diagnostics, identifying biomarkers as targets, and therapeutics, i.e., molecules targeting particular biomarkers, need to develop hand in hand. In this book, the focus has been on therapeutics only, rather than diagnostics.

4. In Europe, Herceptin is no longer patented since 2014 and in the United States since 2019. Since then, other companies have successfully developed biosimilars, which are now sold and used around the globe. For Roche, *trastuzumab* is an "aging blockbuster" by now. On the economics of new oncology drug development see, e.g., DiMasi/Grabowski (2007) and Adams/Brantner (2006).

5. Research views PARP inhibitors (poly(ADP-ribose) polymerase) as a potential treatment for TNBC. A type of targeted therapy, PARP inhibitors block DNA repair and may cause cancer cells to die. In January 2011, the PARP inhibitor *iniparib* (Sanofi) unexpectedly failed Phase III trials. Commentators pointed to the too-broad population in clinical trials for the specific mutation PARP inhibitors work on (Ledford 2012). By the end of 2013, PARP inhibitors "bounced back" (Garber 2013); several PARP inhibitors are in development for breast cancer.

6. *Pertuzumab* (Perjeta): Like *trastuzumab*, *pertuzumab* is a monoclonal antibody that attaches to the HER2 protein. It has also been developed by Genentech/Roche. *Pertuzumab* received FDA approval in 2012 (EMA 2013). Research indicates that it targets a different part of the protein than *trastuzumab* does. This drug is used to treat advanced breast cancer.

7. *Margetuximab* (Margenza) is a mAb with increased ability to bind to HER2+ cells. Developed by MacroGenics, it received FDA approval in 2020.

8. *Ado-trastuzumab emtansine* (Kadcyla) is a mAb antibody-drug conjugate consisting of *trastuzumab* and a chemotherapeutic agent. Developed by Genentech/Roche, it received FDA and EMA approval in 2013.

9. *Fam-trastuzumab deruxtecan* (Enhertu) is a mAb, a chemo agent, and a compound that binds them together. Developed by Daiichi Sankyo/AstraZeneca, it received FDA approval in 2021 and EMA approval in 2021.

10. *Naratinib* (Nerlynx) blocks cells' ability to receive growth signals (EGFR inhibitor). It was approved by the FDA in 2020 and in 2021 by the EMA. It was developed by Puma Biotech.

11. *Tucatinib* (Tukysa) also inhibits HER2+ cell growth. Developed by Seagen/Pfizer, it was approved by the FDA in 2020 and in 2021 by the EMA.

12. *Palbociclib* (Ibrance), developed by Pfizer and FDA approved in 2015, is the first cyclin-dependent kinase 4/6 (CDK4/6) inhibitor. *Palbociclib* works by stopping cancer cells from dividing and growing. For its own tumultuous story of development, see Garber (2014).

13. *Ribociclib* (Kisqali/Kryxana) is also a CDK4/6 inhibitor. Developed by Astex Pharmaceuticals/Novartis, it was FDA and EMA approved in 2017.

14. *Abemaciclib* (Verzenio/Verzenios/Virginio) is also a CDK4/6 inhibitor. Developed by Eli Lilly, it was approved in 2017 by the FDA and in 2018 by the EMA.

15. *Everolimus* (Afinitor): *Everolimus* is a small molecule that blocks mTOR, a protein in cells that normally promotes their growth and division. Together with PI3K and AKT, mTOR is part of the intracellular signaling pathway that controls most hallmarks of cancer, such as cell cycle, survival, metabolism, mobility, and genomic instability (Hanahan/Weinberg 2011). By blocking this protein, *everolimus* can help stop cancer cells and blood vessels from growing. Developed by biotechs and Novartis, it received first FDA approval in 2012.

16. *Alpelisib* (Piqray) was developed by Novartis to work on hormone-receptor-positive, HER2-negative breast cancer with a PIK3CA mutation. It has received approval by the FDA in 2019, and in 2020 by the EMA.

17. *Olaparib* (Lynparza) inhibits the enzyme PARP, which BRCA-mutated cells rely on for cell division. Developed by KuDOS/AstraZeneca, it received approval in 2014 (FDA and EMA).

18. *Talazoparib* (Talzenna) works similar to *olaparib* as a PARP inhibitor. It received approval in 2018 (FDA) and 2019 (EMA) and was developed by Pfizer.

19. *Pembrolizumab* (Keytruda) is a PD1 inhibitor. It was developed by Merck and is the first FDA-approved, tissue-agnostic treatment for solid tumors with certain genetic mutations. It received approval for triple-negative breast cancer in 2021.

20. *Sacituzumab govitecan* (Trodelvy) is a mAb targeting the antigen TROP2. It received FDA and EMA approval in 2021 and was developed by Immunomedics.

21. An immunotherapeutic treatment, the PDL1 inhibitor *atezolizumab* (Tecentriq, developed by Roche) was granted accelerated FDA approval in 2019, yet in summer 2021 Roche withdrew the treatment for triple-negative breast cancer.

Appendix A

1. Sources are several biotech publications, industry and investor news, wire reports, the *New York Times*, *Financial Times*, *Wall Street Journal*, and British newspapers, e.g., the *Evening Standard* and *Daily Mail*.

2. This means that I only considered publications in English. As research on biotech directories, magazines such as *Nature*'s *Bioentrepreneur* and *Biotechnology*, as well as market research reports of the early years, show (e.g., Ernst & Young 1990, 1995) that most commercial molecular cancer therapy research was taking place either in the United States or United Kingdom.

3. In my data source selection, I tried to be most inclusive and chose sources that were available over long periods of time and relevant for the biotech actors in the field themselves. However, I had to make some compromises. For example, the *Financial Times* was only available to me from 1989 to 2003. Reports from business reports in digitized format were available beginning with 1992. Web of Science collection took place in spring 2011, delimiting the data set to the end of 2010. LexisNexis data collection as well as their analyses took place while I was a visiting scholar at the Center for European Studies, Harvard University, spring 2012.

Appendix B

1. This is a Java-run software, written in the programming language Scala. See http://nlp.stanford.edu/software/tmt/tmt-0.4/ for further details on the toolkit.

Bibliography

Abbott, Andrew. 1988. "Transcending General Linear Reality." *Sociological Theory* 6: 169–188.
Abbott, Andrew. 1992. "From Causes to Events: Notes on Narrative Positivism." *Sociological Methods and Research* 20: 428–455.
Abbott, Andrew. 2007a. "Against Narrative: A Preface to Lyrical Sociology." *Sociological Theory* 25: 67–99.
Abbott, Andrew. 2007b. "Mechanisms and Relations." *Sociologica* 2, http://www.sociologica.mulino.it/doi/10.2383/24750.
Abbott, Andrew/Alexandra Hrycak. 1990. "Measuring Resemblance in Sequence Data: An Optimal Matching Analysis of Musicians' Careers." *American Journal of Sociology* 96: 144–185.
Abolafia, Mitchel Y. 1996. *Making Markets: Opportunism and Restraint on Wall Street*. Cambridge, MA: Harvard University Press.
Abolafia, Mitchel Y. 2010. "Narrative Construction as Sensemaking: How a Central Bank Thinks." *Organization Studies* 31: 349–367.
Adam, Barbara/Ulrich Beck/Joost van Loon. 2000. *The Risk Society and Beyond: Critical Issues for Social Theory*. London; Thousand Oaks: SAGE.
Adams, Christopher P./Van V. Brantner. 2006. "Estimating the Cost of New Drug Development: Is It Really $802 Million?" *Health Affairs* 25: 420–428.
Aggarwal, Saurabh. 2010. "Targeted Cancer Therapies." *Nature Reviews Drug Discovery* 9: 427–428.
Agus, David B./Elizabeth M. Jaffee/Chi Van Dang. 2021. "Cancer Moonshot 2.0." *The Lancet Oncology* 22: 164–165.
Aisenbrey, Silke/Anette Fasang. 2017. "The Interplay of Work and Family Trajectories over the Life Course: Germany and the United States in Comparison." *American Journal of Sociology* 122: 1448–1484.
Akerlof, George A. 1970. "The Market for Lemons." *Qualitative Journal of Economics* 84: 488–500.
Akerlof, George A./Robert J. Shiller. 2009. *Animal Spirits. How Human Psychology Drives the Economy, and Why It Matters for Global Capitalism*. Princeton, NJ: Princeton University Press.

Akrich, Madeleine/Michel Callon/Bruno Latour. 2002. "The Key to Success in Innovation, Part I: The Art of Interessement." *International Journal of Innovation Management* 6: 187–206.

Alexander, Andrew. 1995. "New Cancer Drug Sends British Biotech Soaring." *Daily Mail*, December 1, 1995, 69.

Alexander, Jeffrey C/Bernhard Giesen/Richard Münch/Neil J. Smelser (eds.). 1987. *The Micro-Macro Link*. Berkeley: University of California Press.

Allansdottir, Agnes/Andrea Bonaccorsi/Alfonso Gambardella/Myriam Mariani/Luigi Orsenigo/Fabio Pammolli/Massimo Riccaboni. 2002. *Innovation and Competitiveness in European Biotechnology*. Luxembourg: Office for Official Publications of the European Communities.

Allison, Malorye. 2008. "Is Personalized Medicine Finally Arriving?" *Nature Biotechnology* 26: 509–517.

Almeida, Paul/Bruce Kogut. 1999. "Localization of Knowledge and the Mobility of Engineers in Regional Networks." *Management Science* 47: 905–917.

Alvesson, Mats/Dan Kärreman. 2000. "Varieties of Discourse: On the Study of Organizations through Discourse Analysis." *Human Relations* 53: 1125–1149.

Amir-Aslani, Arsia/Vincent Mangematin. 2010. "The Future of Drug Discovery and Development: Shifting Emphasis Towards Personalized Medicine." *Technological Forecasting & Social Change* 77: 208–217.

Angermuller, Johannes/Dominique Maingueneau/Ruth Wodak. 2014. "The Discourse Studies Reader: An Introduction," in Johannes Angermuller/Dominique Maingueneau/Ruth Wodak (eds.), *The Discourse Studies Reader: Main Currents in Theory and Analysis*. Amsterdam: John Benjamins: 1–14.

Anonymous. 1996. "Letters to the Editor." *The Lancet* 348: 263.

Anonymous. 2005. "King in the Kingdom of Uncertainty." *Nature Biotechnology* 23: 1025–1025.

Arora, Ashish/Alfonso Gambardella. 1994. "Evaluating Technological Information and Using It: Scientific Knowledge, Technological Capability, and External Linkages in Biotechnology." *Journal of Economic Behavior and Organization* 24: 91–114.

Aspers, Patrik. 2006. *Markets in Fashion*. Oxford: Routledge.

Aspers, Patrik. 2009. "Knowledge and Valuation in Markets." *Theory and Society* 38: 111–131.

Aspers, Patrik. 2010. *Orderly Fashion. A Sociology of Markets*. Princeton, NJ: Princeton University Press.

Aspers, Patrik. 2011. *Markets*. Cambridge, UK: Polity Press.

Aspinall, Mara G./Richard G. Hamermesh. 2007. "Realizing the Promise of Personalized Medicine." *Harvard Business Review* 85: 108–117.

Atkins, Joshua H./Leland J. Gershell. 2002. "Selective Anticancer Drugs." *Nature Reviews Drug Discovery* 1: 491–492.

Bail, Christopher A. 2014. "The Cultural Environment: Measuring Culture with Big Data." *Theory and Society* 43: 465–482.

Bail, Christopher A. 2016a. "Cultural Carrying Capacity: Organ Donation Advocacy, Discursive Framing, and Social Media Engagement." *Social Science & Medicine* 165: 280–288.

Bail, Christopher A. 2016b. *Terrified. How Anti-Muslim Fringe Organizations Became Mainstream*. Princeton, NJ: Princeton University Press.

Baker, Monya. 2005. "In Biomarkers We Trust?" *Nature Biotechnology* 23: 297–304.

Baker, Wayne E. 1984. "The Social Structure of a National Securities Market." *American Journal of Sociology* 89: 775–811.

Bandelj, Nina. 2012. "Relational Work and Economic Sociology." *Politics & Society* 40: 175–201.

Bartel, Caroline A./Raghu Garud. 2009. "The Role of Narratives in Sustaining Organizational Innovation." *Organization Science* 20: 107–117.

Bartholomew, Susan. 1997. "National Systems of Biotechnology Innovation: Complex Interdependence in the Global System." *Journal of International Business Studies* 28: 241–266.

Baselga, Jose/Debasish Tripathy/John Mendelsohn/Sharon Baughman/Christopher C Benz/Lucy Dantis/Nancy T Sklarin/Andrew D Seidman/Clifford A Hudis/Jackie Moore. 1996. "Phase II Study of Weekly Intravenous Recombinant Humanized Anti-p185HER2 Monoclonal Antibody in Patients with HER2/neu-overexpressing Metastatic Breast Cancer." *Journal of Clinical Oncology* 14: 737–744.

Basov, Nikita/Ronald Breiger/Iina Hellsten. 2020. "Socio-semantic and Other Dualities." *Poetics* 78.

Bauer, Martin W./George Gaskell. 2002. *Biotechnology: The Making of a Global Controversy*. Cambridge: Cambridge University Press.

Bazell, Robert. 1998. *Her-2: The Making of Herceptin, a Revolutionary Treatment for Breast Cancer*. New York: Random House.

Bearman, Peter S. 2015. "Big Data and Historical Social Science." *Big Data & Society* 2, http://bds.sagepub.com/content/2/2/2053951715612497.full.

Bearman, Peter S./Robert Faris/James Moody. 1999. "Blocking the Future." *Social Science History* 23: 501–533.

Bearman, Peter S./Katherine Stovel. 2000. "Becoming a Nazi: A Model for Narrative Networks." *Poetics* 27: 69–90.

Beckert, Jens. 2010. "How Do Fields Change? The Interrelations of Institutions, Networks, and Cognition in the Dynamics of Markets." *Organization Studies* 31: 605–627.

Beckert, Jens. 2013. "Imagined Futures: Fictional Expectations in the Economy." *Theory and Society* 42: 219–240.

Beckert, Jens. 2016. *Imagined Futures. Fictional Expectations and Capitalist Dynamics*. Cambridge, MA: Harvard University Press.

Beckert, Jens. 2019. "Markets from Meaning: Quality Uncertainty and the Intersubjective Construction of Value." *Cambridge Journal of Economics* 44: 285–301.

Beckert, Jens. 2021. "The Firm as an Engine of Imagination: Organizational Prospection and the Making of Economic Futures." *Organization Theory* 2: 1–21.

Beckert, Jens/Patrik Aspers (eds.). 2011. *The Worth of Goods: Valuation and Pricing in the Economy*. Oxford: Oxford University Press.

Beckert, Jens/Richard Bronk (eds.). 2018. *Uncertain Futures: Imaginaries, Narratives, and Calculation in the Economy*. Oxford: Oxford University Press.

Bell, Andrew. 1997. "Biomira Faces Long Road to Recovery." *The Globe and Mail*, December 5, 1997.

Bell, Kirsten. 2013. "Biomarkers, the Molecular Gaze and the Transformation of Cancer Survivorship." *BioSocieties* 8: 124–143.

Benjamin, Beth/Joel M. Podolny. 1999. "Status, Quality, and Social Order in the California Wine Industry." *Administrative Science Quarterly* 44: 563–589.

Berenson, Alex. 2005. "Blockbuster Drugs Are So Last Century." *The New York Times*, July 3, 2005, Business Section, p. 1.

Beunza, Daniel/David Stark. 2012. "From Dissonance to Resonance: Cognitive Interdependence in Quantitative Finance." *Economy and Society* 41: 383–417.

Biggart, Nicole Woolsey/Thomas D. Beamish. 2003. "The Economic Sociology of Conventions: Habit, Custom, Practice, and Routine in Market Order." *Annual Review of Sociology* 29: 443–464.

Bijker, Wiebe E./Thomas Hughes/Trevor Pinch. 1989. *The Social Construction of Technological Systems*. Boston: MIT Press.

Bingham, Christopher B./Steven J. Kahl. 2013. "The Process of Schema Emergence: Assimilation, Deconstruction, Unitization and the Plurality of Analogies." *Academy of Management Journal* 56: 14–34.

Biotechnology Business News. 1995a. "Biomira Releases Interim Data on Anticancer Vaccine." *Biotechnology Business News*, March 15, 1995.

Biotechnology Business News. 1995b. "Sugen—Building a Cancer Franchise." *Biotechnology Business News*, October 18, 1995.

Birch, Steve. 2002. *The Cancer Outlook to 2007*. London: Business Insights.

Blackstone, Tim. 1996. "Biotech Star's Testing Time on Cancer Claim." *Mail on Sunday*, 13. Oktober 1996, 48.

Blair-Loy, Mary. 1999. "Career Patterns of Executive Women in Finance: An Optimal Matching Analysis." *American Journal of Sociology* 104: 1346–1397.

Blei, David M. 2012. "Probabilistic Topic Models." *Communications of the ACM* 55: 77–84.

Blei, David M./John D. Lafferty. 2009. "Topic Models," in Ashok Srivastava/Mehran Sahami (eds.), *Text Mining: Classification, Clustering, and Applications*. Boca Raton, FL: Chapman & Hall: 71–93.

Blei, David M./Andrew Y. Ng/Michael Jordan. 2003. "Latent Dirichlet Allocation." *Journal of Machine Learning Research* 3: 993–1022.

Blondel, Vincent D./Jean-Loup Guillaume/Renaud Lambiotte/Etienne Lefebvre. 2008. "Fast Unfolding of Communities in Large Networks." *Journal of Statistical Mechanics: Theory and Experiment* 10: P10008–P10019.

Boje, David M. 1991. "The Storytelling Organization: A Study of Story Performance in an Office-Supply Firm." *Administrative Science Quarterly* 36: 106–216.

Boje, David M. 2001. *Narrative Methods for Organizational and Communication Research*. London; Thousand Oaks: SAGE.
Boltanski, Luc. 1996. *Endless Disputes from Intimate Injuries to Public Denunciations*. Cornell University, Department of Sociology.
Boltanski, Luc/Laurent Thévenot. 1999. "The Sociology of Critical Capacity." *European Journal of Social Theory* 2: 359–377.
Boltanski, Luc/Laurent Thévenot. 2006. *On Justification: Economies of Worth*. Princeton, NJ: Princeton University Press.
Booth, Bruce. 2012. "Cancer Drug Targets: The March of the Lemmings." *Forbes*. June 7, 2012, http://onforb.es/LzE7MX.
Booth, Bruce/Robert Glassman/Philip Ma. 2003. "Oncology's Trials." *Nature Reviews Drug Discovery* 2: 609–610.
Booth, Bruce/Rodney Zemmel. 2004. "Prospects for Productivity." *Nature Reviews Drug Discovery* 3: 451–456.
Borup, Mads/Nik Brown/Kornelia Konrad/Harro van Lente. 2006. "The Sociology of Expectations in Science and Technology." *Technology Analysis & Strategic Management* 18: 285–298.
Bothner, Matthew S./Young-Kyu Kim/Edward Bishop Smith. 2012. "How Does Status Affect Performance? Status as an Asset vs. Status as a Liability in the PGA and NASCAR." *Organization Science* 23: 416–433.
Bourret, Pascale. 2005. "BRCA Patients and Clinical Collectives: New Configurations of Action in Cancer Genetics Practices." *Social Studies of Science* 35: 41–68.
Bourret, Pascale/Peter Keating/Alberto Cambrosio. 2014. "From BRCA to BRCAness," in Sahra Gibbon/Galen Joseph/Jessica Mozersky/Andrea zur Nieden/Sonja Palfner (eds.), *Breast Cancer Gene Research and Medical Practices: Transnational Perspectives in the Time of BRCA*. Oxon: Routledge: 175–193.
Bourret, Pascale/Andrei Mogoutov/Claire Julian-Reynier/Alberto Cambrosio. 2006. "A New Clinical Collective for French Cancer Genetics: A Heterogeneous Mapping Analysis." *Science Technology Human Values* 31: 431–464.
Bowker, Geoffrey/Susan Leigh Star. 1999. *Sorting Things Out*. Cambridge, MA: MIT Press.
Boyack, Kevin W./Richard Klavans/Katy Börner. 2005. "Mapping the Backbone of Science." *Scientometrics* 64: 351–374.
Bragd, Annica/Dorit Christensen/Barbara Czarniawska/Maria Tullberg. 2008. "Discourse as the Means of Community Creation." *Scandinavian Journal of Management* 24: 199–208.
Breiger, Ronald L. 1974. "The Duality of Persons and Groups." *Social Forces* 53: 181–190.
Breiger, Ronald L. 1976. "Career Attributes and Network Structure: A Blockmodel Study of a Biomedical Research Specialty." *American Sociological Review* 41: 117–135.
Breiger, Ronald L. 2005. "Culture and Classification in Markets: An Introduction." *Poetics* 33: 157–162.
Breiger, Ronald L./Robin Wagner-Pacifici/John W. Mohr. 2018. "Capturing Distinctions While Mining Text Data: Toward Low-tech Formalization for Text Analysis." *Poetics* 68: 104–119.

Breschi, Stefano/Franco Malerba/Luigi Orsenigo. 2000. "Technological Regimes and Schumpeterian Patterns of Innovation." *Economic Journal*: 388–410.

Brower, Vicki. 1999. "Tumor Angiogenesis—New Drugs on the Block." *Nature Biotechnology* 17: 963–968.

Brown, John Seely/Stephen Denning/Katalina Groh/Laurence Prusak (eds.). 2005. *Storytelling in Organizations: Why Storytelling Is Transforming 21st Century Organizations and Management*. Burlington, MA: Elsevier Butterworth-Heinemann.

Brown, Nik/Mike Michael. 2003. "A Sociology of Expectations: Retrospecting Prospects and Prospecting Retrospects." *Technology Analysis & Strategic Management* 15: 3–18.

Brown, Nik/Brian Rappert/Andrew Webster (eds.). 2000. *Contested Futures: A Sociology of Prospective Techno-Science*. Burlington, VT: Ashgate.

Buonansegna, Erika/Søren Salomo/Anja M. Maier/Jason Li-Ying. 2014. "Pharmaceutical New Product Development: Why Do Clinical Trials Fail?" *R&D Management* 44: 189–202.

Burt, Ronald S. 1992a. "The Social Structure of Competition," in Nitin Nohria/Robert Eccles (eds.), *Networks and Organization*. Boston: Harvard Business School Press: 57–91.

Burt, Ronald S. 1992b. *Structural Holes: The Social Structure of Competition*. Cambridge, MA: Harvard University Press.

Business Insights. 2001. *The Blockbuster Drug Outlook*. London: Business Insights.

Business Insights. 2011. *The Cancer Market Outlook to 2016*. London: Business Insights.

Business Wire. 1990. "Roche Group Earnings Up 33 Percent to New Record, Annual Dividend of Sfr. 38 Proposed." Business Wire, April 24, 1990.

Business Wire. 1991. "Genentech to Test Potential Therapy for Breast and Ovarian Cancers." Business Wire, April 5, 1991.

Business Wire. 1992. "Sugen, Inc. Raises $5 Million in Private Financin." Business Wire, October 15, 1992.

Business Wire. 1993. "Medarex Announces Initiation of Human Clinical Trial for New Breast and Ovarian Cancer Therapeutic." Business Wire, May 20, 1993.

Business Wire. 1994a. "British Bio-technology Announces Major Fevelopments." Business Wire, June 16, 1994.

Business Wire. 1994b. "Genentech Reports 1993 Fourth Quarter and Year End Results." Business Wire, January 19, 1994.

Business Wire. 1995. "British Biotech Batimastat Clinical Program Revised." Business Wire, February 17, 1995.

Business Wire. 1997a. "Genentech Announces Phase III Investigational Trial Results for HER2 Monoclonal Antibody in Metastatic Breast Cancer." Business Wire, December 19, 1997.

Business Wire. 1997b. "Genentech Focuses on Cancer as It Launches BioOncology Initiative." Business Wire, May 15, 1997.

Business Wire. 1998a. "Biotechnology Breakthrough in Breast Cancer Wins FDA Approval." Business Wire, September 25, 1998.

Business Wire. 1998b. "Genentech Submits Application for FDA Approval of HER2 Antibody, Herceptin." *Business Wire*, May 4, 1998.

Çalışkan, Koray. 2010. *Market Threads: How Cotton Farmers and Traders Create a Global Commodity*. Princeton, NJ: Princeton University Press.

Çalışkan, Koray/Michel Callon. 2010. "Economization, Part 2: A Research Programme for the Study of Markets." *Economy and Society* 39: 1–32.

Callon, Michel. 1986a. "The Sociology of an Actor-Network: The Case of the Electric Vehicle," in Michel Callon/John Law/A. Rip (eds.), *Mapping the Dynamics of Science and Technology: Sociology of Science in the Real World*. Basingstoke, UK: Macmillan Press: 19–34.

Callon, Michel. 1986b. "Some Elements of a Sociology of Translation: Domestication of the Scallops and the Fisherman of St. Brieuc Bay," in John Law (ed.), *Power, Action, and Belief: A New Sociology of Knowledge?* London, Boston, and Henley: Routledge & Kegan Paul: 196–233.

Callon, Michel. 1992. "The Dynamics of Techno-Economic Networks," in Rod Coombs/Paolo Saviotti/Vivien Walsh (eds.), *Technological Change and Company Strategies*. London: Academic Press: 72–102.

Callon, Michel. 1998a. "Introduction: The Embeddedness of Economic Markets in Economics," in Michel Callon (ed.), *The Laws of the Markets*. Oxford: Blackwell: 1–57.

Callon, Michel (ed.). 1998b. *The Laws of the Markets*. Oxford, UK: Blackwell.

Callon, Michel. 2006. "Can Methods for Analysing Large Numbers Organize a Productive Dialogue with the Actors They Study?" *European Management Review* 3: 7–16.

Callon, Michel. 2007a. "An Essay on the Growing Contribution of Economic Markets to the Proliferation of the Social." *Theory, Culture & Society* 24: 139–163.

Callon, Michel. 2007b. "What Does It Mean to Say That Economics Is Performative?," in Donald MacKenzie/Fabian Muniesa/Lucia Siu (eds.), *Do Economists Make Markets? On the Performativity of Economics*. Princeton, NJ: Princeton University Press: 311–357.

Callon, Michel. 2008. "Economic Markets and the Rise of Interactive Agencements: From Prosthetic Agencies to Habilitated Agencies," in Trevor Pinch/Richard Swedberg (eds.), *Living in a Material World: Economic Sociology Meets Science and Technology Studies*. Cambridge, MA: MIT Press: 29–56.

Callon, Michel. 2015. "Revisiting Marketization: From Interface-Markets to Market-Agencements." *Consumption Markets & Culture*: 1–21.

Callon, Michel. 2021. *Markets in the Making: Rethinking Competition, Goods, and Innovation*. New York: Zone Books.

Callon, Michel/Jean-Pierre Courtial/William A. Turner/Serge Bauin. 1983. "From Translations to Problematic Networks: An Introduction to Co-word Analysis." *Social Science Information* 22: 191–235.

Callon, Michel/John Law/Arie Rip. 1986. *Mapping the Dynamics of Science and Technology: Sociology of Science in the Real World*. Houndsmill: Macmillan Press.

Callon, Michel/Fabian Muniesa. 2005. "Economic Markets as Calculative Collective Devices." *Organization Studies* 26: 1229–1250.

Cambrosio, Alberto/Pascal Cottereau/Stefan Popowycz/Andrei Mogoutov/Tania Vichnevskaia. 2013. "Analysis of Heterogenous Networks: The ReseauLu Project," in Bernard Reber/Claire Brossaud (eds.), *Digital Cognitive Technologies: Epistemology and the Knowledge Economy*. Hoboken, NJ: John Wiley & Sons, Inc.: 137–152.

Cambrosio, Alberto/Peter Keating. 1995. *Exquisite Specificity: The Monoclonal Antibody Revolution*. New York: Oxford University Press.

Cambrosio, Alberto/Peter Keating/Simon Mercier/Grant Lewison/Andrei Mogoutov. 2006. "Mapping the Emergence and Development of Translational Cancer Research." *European Journal of Cancer* 42: 3140–3148.

Cambrosio, Alberto/Peter Keating/Andrei Mogoutov. 2004. "Mapping Collaborative Work and Innovation in Biomedicine: A Computer-Assisted Analysis of Antibody Reagent Workshops." *Social Studies of Science* 34: 325–364.

Campeau, Philippe M./William D. Foulkes/Marc D. Tischkowitz. 2008. "Hereditary Breast Cancer: New Genetic Developments, New Therapeutic Avenues." *Human Genetics* 124: 31–42.

Campisi, Judith. 2005. "Suppressing Cancer: The Importance of Being Senescent." *Science* 309: 886–887.

Cancer Genome Atlas Network. 2012. "Comprehensive Molecular Portraits of Human Breast Tumours." *Nature* 490: 61–70.

Cantor, David. 2007. "Introduction: Cancer Control and Prevention in the Twentieth Century." *Bulletin of the History of Medicine* 81: 1–38.

Cantor, David (ed.). 2008. *Cancer in the Twentieth Century*. Baltimore: Johns Hopkins University Press

Carley, Kathleen M. 1997. "Network Text Analysis: The Network Position of Concepts," in Carl W. Roberts (ed.), *Text Analysis for the Social Sciences*. Hillsdale, NJ: Lawrence Erlbaum Associates: 79–100.

Carney, Walter P. 2005. "HER2 Status Is an Important Biomarker in Guiding Personalized HER2 Therapy." *Personalized Medicine* 2: 317–324.

Casper, Steven. 2007. "How Do Technology Clusters Emerge and Become Sustainable? Social Network Formation and Inter-firm Mobility within the San Diego Biotechnology Cluster." *Research Policy* 36: 438–455.

Casper, Steven/Catherine Matraves. 2003. "Institutional Frameworks and Innovation in the German and UK Pharmaceutical Industry." *Research Policy* 32: 1865–1879.

Casper, Steven/Fiona Murray. 2005. "Careers and Clusters: Analyzing the Career Network Dynamic of Biotechnology Clusters." *Journal of Engineering and Technology Management* 22: 51–74.

Cattani, Gino/Joseph Porac/Howard Thomas. 2017. "Categories and Competition." *Strategic Management Journal* 38: 64–92.

Chabner, Bruce A./Thomas G. Roberts. 2005. "Chemotherapy and the War on Cancer." *Nature Reviews Cancer* 5: 65–72.

Charon, Rita. 2006. *Narrative Medicine: Honoring the Stories of Illness*. Oxford; New York: Oxford University Press.

Chase, Marilyn. 1992. "Genentech Drug to Treat Cancer Appears to be Safe." *Wall Street Journal*, May 22, 1992.

Chavalarias, David/Jean-Philippe Cointet. 2013. "Phylomemetic Patterns in Science Evolution—The Rise and Fall of Scientific Fields." *PloS One* 8: e54847.

Chavalarias, David/Jean-Philippe Cointet/Lise Cornilleau/Tam Kien Duong/Andreï Mogoutov/Lionel Villard/Camille Roth/Thierry Savy. 2011. "Streams of Media Issues, Monitoring World Food Security." http://pulseweb.cortext.net/static/files/wp1.pdf.

Chin, Lynda/Jannik N. Andersen/P. Andrew Futreal. 2011. "Cancer Genomics: From Discovery Science to Personalized Medicine." *Nature Medicine* 17: 297–303.

Christie, Bryan. 1994. "New Drug Offers Hope to Cancer Patients." *The Scotsman*, March 18, 1994.

Cobleigh, Melody A./Charles L. Vogel/Debu Tripathy/Nicholas J. Robert/Susy Scholl/Louis Fehrenbacher/Janet M. Wolter/Virginia Paton/Steven Shak/Gracie Lieberman/Dennis J. Slamon. 1999. "Multinational Study of the Efficacy and Safety of Humanized Anti-HER2 Monoclonal Antibody in Women Who Have HER2-Overexpressing Metastatic Breast Cancer That Has Progressed After Chemotherapy for Metastatic Disease." *Journal of Clinical Oncology* 17: 2639.

Cooren, François. 2004. "Textual Agency: How Texts Do Things in Organizational Settings." *Organization* 11: 373–393.

Cornelissen, Joep P./Jean S. Clarke. 2010. "Imagining and Rationalizing Opportunities: Inductive Reasoning, and the Creation and Justification of New Ventures." *Academy of Management Review* 35: 539–557.

Coussens, Lisa M./Barbara Fingleton/Lynn M. Matrisian. 2002. "Matrix Metalloproteinase Inhibitors and Cancer—Trials and Tribulations." *Science* 295: 2387–2392.

Coussens, Lisa/Teresa L. Yang-Feng/Yu-Cheng Liao/Chen Ellson/Alane Gray/John McGrath/Peter H. Seeburg/Towia A. Libermann/Joseph Schlessinger/Uta Francke/Arthur Levinson/Axel Ullrich. 1985. "Tyrosine Kinase Receptor with Extensive Homology to EGF Receptor Shares Chromosomal Location with neu Oncogene." *Science* 230: 1132–1139.

Couzin-Frankel, Jennifer/Yasmin Ogale. 2011. "Once on 'Fast Track,' Avastin Now Derailed." *Science* 333: 143–144.

Cowley, Geoffrey. 1993. "Immunotherapy." *The Globe and Mail*, October 28, 1993.

Crossley, Nick. 2010. *Towards Relational Sociology*. Oxford: Routledge.

Crossley, Nick. 2015. "Relational Sociology and Culture: A Preliminary Framework." *International Review of Sociology* 25: 65–85.

Cui, Weiwei/Shixia Liu/Li Tan/Conglei Shi/Yangqiu Song/Zekai J Gao/Huamin Qu/Xin Tong. 2011. "Textflow: Towards Better Understanding of Evolving Topics in Text." *IEEE Transactions on Visualization and Computer Graphics* 17: 2412–2421.

Czarniawska, Barbara. 1997. *Narrating the Organization: Dramas of Institutional Identity*. Chicago: University of Chicago Press.

Czarniawska, Barbara. 1998. *A Narrative Approach to Organization Studies*. Thousand Oaks, CA: SAGE.

Czarniawska, Barbara. 2004. *Narratives in Social Science Research*. London: SAGE.
Dabrowski, Wojtek. 2003. "Biomira Falls 63% on Failed Vaccine Trial." *National Post's Financial Post & FP Investing*, June 18, 2003.
Daily Mail. 1992. "Breast Cancer 'Wonder Drug' on Trial." *Daily Mail*, June 16, 1992.
Daily Mail. 1995. "Ex-Biotech Boffin a Hit at Chiro." *Daily Mail*, January 18, 1995, 57.
Davidson, Sylvia. 1996. "Is British Biotech's Marimastat a Major Cancer Drug?" *Nature Biotechnology* 14: 819–820.
de Vaan, Mathijs/Balazs Vedres/David Stark. 2015. "Game Changer: The Topology of Creativity." *American Journal of Sociology* 120: 1144–1194.
Denning, Stephen. 2005. *The Leader's Guide to Storytelling: Mastering the Art and Discipline of Business Narrative*. San Francisco: Jossey-Bass.
Dépelteau, François. 2015. "Relational Sociology, Pragmatism, Transactions and Social Fields." *International Review of Sociology* 25: 45–64.
Deuten, J. Jasper/Arie Rip. 2000. "Narrative Infrastructure in Product Creation Processes." *Organization* 7: 69–93.
DeVita, Vincent T. 2015. *The Death of Cancer*. With Elizabeth DeVita-Raeburn. New York: Farrar, Straus & Giroux.
DeVita, Vincent T./Edward Chu. 2008. "A History of Cancer Chemotherapy." *Cancer Research* 68: 8643–8653.
Diaz-Bone, Rainer. 2011. "The Methodological Standpoint of the 'économie des conventions.'" *Historical Social Research* 36: 43–63.
DiMaggio, Paul. 1997. "Culture and Cognition." *Annual Review of Sociology* 23: 263–287.
DiMaggio, Paul. 2011. "Cultural Networks," in John Scott/Peter J. Carrington (eds.), *The SAGE Handbook of Social Network Analysis*. London: SAGE: 286–300.
DiMaggio, Paul/Manish Nag/David Blei. 2013. "Exploiting Affinities between Topic Modeling and the Sociological Perspective on Culture: Application to Newspaper Coverage of U.S. Government Arts Funding." *Poetics* 41: 570–606.
DiMaggio, Paul/Walter W. Powell. 1983. "The Iron Cage Revisited: Institutional Isomorphism and Collective Rationality." *American Sociological Review* 48: 147–160.
DiMaggio, Paul/Walter W. Powell. 1991. "Introduction," in Walter Powell/Paul DiMaggio (eds.), *The New Institutionalism in Organizational Analysis*. Chicago: University of Chicago Press: 1–38.
DiMasi, Joseph A./Henry G. Grabowski. 2007. "Economics of New Oncology Drug Development." *Journal of Clinical Oncology* 25: 209–216.
Donati, Pierpaolo. 2010. *Relational Sociology: A New Paradigm for the Social Sciences*. Oxford: Routledge.
Donati, Pierpaolo. 2015. "Manifesto for a Critical Realist Relational Sociology." *International Review of Sociology* 25: 86–109.
Dooley, Joseph. 2002. *New Cancer Therapeutics*. London: Business Insights.
Dougherty, Deborah/Danielle D. Dunne. 2011. "Organizing Ecologies of Complex Innovation." *Organization Science* 22: 1214–1223.
Dow, Alistair. 1992. "Biomira: Sexy Stock with High Hopes." *The Toronto Star*, January 11, 1992, C2.

Downing, Nicholas S./Jenerius A. Aminawung/Nilay D. Shah/Joel B. Braunstein/Harlan M. Krumholz/Joseph S. Ross. 2012. "Regulatory Review of Novel Therapeutics—Comparison of Three Regulatory Agencies." *New England Journal of Medicine* 366: 2284–2293.
Drews, Jürgen. 2004. "Paul Ehrlich: Magister Mundi." *Nature Reviews Drug Discovery* 3: 797–801.
Drews, Jürgen/Stefan Ryser. 1997. "Drug Development: The Role of Innovation in Drug Development." *Nature Biotechnology* 15: 1318–1319.
Durand, Rodolphe/Nina Granqvist/Anna Tyllström. 2017. "From Categories to Categorization: A Social Perspective on Market Categorization," in Rodolphe Durand/Nina Granqvist/Anna Tyllström (eds.), *From Categories to Categorization: Studies in Sociology, Organizations and Strategy at the Crossroads*. Bingley: Emerald Publishing: 51: 3–30.
Durand, Rodolphe/Mukti Khaire. 2017. "Where Do Market Categories Come from and How? Distinguishing Category Creation from Category Emergence." *Journal of Management* 43: 87–110.
Ebers, Mark/Walter W. Powell. 2007. "Biotechnology: Its Origins, Organization, and Outputs." *Research Policy* 36: 433–437.
Ecker, Dawn M./Susan Dana Jones/Howard L. Levine. 2015. "The Therapeutic Monoclonal Antibody Market." *mAbs* 7: 9–14.
Eckhardt, Bedrich L./Prudence A. Francis/Belinda S. Parker/Robin L. Anderson. 2012. "Strategies for the Discovery and Development of Therapies for Metastatic Breast Cancer." *Nature Reviews Drug Discovery* 11: 479–497.
Edelmann, Achim/John W. Mohr. 2018. "Formal Studies of Culture: Issues, Challenges, and Current Trends." *Poetics* 68: 1–9.
Edelmann, Achim/Tom Wolff/Danielle Montagne/Christopher A. Bail. 2020. "Computational Social Science and Sociology." *Annual Review of Sociology* 46: 61–81.
Elliott, Jane. 2005. *Using Narrative in Social Research: Qualitative and Quantitative Approaches*. London: SAGE.
Ellsworth, Rachel/David Decewicz/Craig Shriver/Darrell Ellsworth. 2010. "Breast Cancer in the Personal Genomics Era." *Current Genomics* 11: 146–161.
Emirbayer, Mustafa. 1997. "Manifesto for a Relational Sociology." *American Journal of Sociology* 103: 281–317.
Emirbayer, Mustafa/Victoria Johnson. 2008. "Bourdieu and Organizational Analysis." *Theory and Society* 37: 1–44.
Emirbayer, Mustafa/Ann Mische. 1998. "What Is Agency?" *American Journal of Sociology* 103: 962–1023.
Ernst & Young. 1990. *European Biotech*. London: Ernst & Young.
Ernst & Young. 1995. *European Biotech 95: Gathering Momentum*. Brussels: Ernst & Young.
Espeland, Wendy/Mitchell L. Stevens. 1998. "Commensuration as a Social Process." *Annual Review of Sociology* 24: 313–343.
Esposito, Elena. 2013. "The Structures of Uncertainty: Performativity and Unpredictability in Economic Operations." *Economy and Society* 42: 102–129.

Evans, James A./Pedro Aceves. 2016. "Machine Translation: Mining Text for Social Theory." *Annual Review of Sociology* 42: 21–50.
Evans, John H. 2002. *Playing God? Human Genetic Engineering and the Rationalization of Public Bioethical Debate*. Chicago: University of Chicago Press.
Evening Standard. 1995. "£70m Drug Trial Blow to Biotech." *Evening Standard*, February 17, 1995.
Extel Examiner. 1996. "Chiroscience Says New Cancer Drug 'Better than British Biotech's Marimastat.'" *Extel Examiner*, 25, April 1996.
Fairclough, Norman. 1995. *Critical Discourse Analysis: The Critical Study of Language*. London: Longman.
Fairclough, Norman. 2005. "Discourse Analysis in Organization Studies: The Case for Critical Realism." *Organization Studies* 26: 915–939.
Fairclough, Norman. 2013. *Critical Discourse Analysis: The Critical Study of Language*. 2nd ed. Oxford: Routledge.
Fairhurst, Gail T./Linda L. Putnam. 2004. "Organizations as Discursive Constructions." *Communication Theory* 14: 5–26.
Favereau, Olivier/Olivier Biencourt/Francois Eymard-Duvernay. 2002. "Where Do Markets Come From? From (Quality) Conventions!," in Olivier Favereau/Emmanuel Lazega (eds.), *Conventions and Structures in Economic Organization*. Cheltenham: Edward Elgar: 213–252.
Favereau, Olivier/Emmanuel Lazega (eds.). 2002. *Conventions and Structures in Economic Organization: Markets, Networks, and Hierarchies*. Cheltenham: Edward Elgar.
Fernandez, Jose-Maria/Roger M. Stein/Andrew W. Lo. 2012. "Commercializing Biomedical Research through Securitization Techniques." *Nature Biotechnology* 30: 964–975.
Ferrara, Napoleone/Kenneth J. Hillan/Hans-Peter Gerber/William Novotny. 2004. "Discovery and Development of Bevacizumab, an Anti-VEGF Antibody for Treating Cancer." *Nature Reviews Drug Discovery* 3: 391–400.
Financial Times. 1994. "British Biotech." *Financial Times*, May 24, 1994.
Fine, Gary Alan/Corey D. Fields. 2008. "Culture and Microsociology: The Anthill and the Veldt." *The Annals of the American Academy of Political and Social Science* 619: 130–148.
Fisher, Lawrence. 1997. "Hope Near the End of the Pipeline." *The New York Times*, May 1, 1997, 1, Business.
Fleming, Lee/Charles King III/Adam I. Juda. 2007. "Small Worlds and Regional Innovation." *Organization Science* 18: 938–954.
Fligstein, Neil. 2001. *The Architecture of Markets: An Economic Sociology of Twenty-First-Century Capitalist Societies*. Princeton, NJ: Princeton University Press.
Fligstein, Neil/Doug McAdam. 2012. *A Theory of Fields*. New York: Oxford University Press.
Fortun, Michael. 2001. "Mediated Speculations in the Genomics Futures Markets." *New Genetics & Society* 20: 139–156.

Foulkes, William D./Ian E. Smith/Jorge S. Reis-Filho. 2010. "Triple-Negative Breast Cancer." *New England Journal of Medicine* 363: 1938–1948.
Fourcade, Marion. 2007. "Theories of Markets and Theories of Society." *American Behavioral Scientist* 50: 1015–1034.
Fourcade, Marion. 2011. "Cents and Sensibility: Economic Valuation and the Nature of 'Nature.'" *American Journal of Sociology* 116: 1721–1777.
Frankfurter Allgemeine Zeitung. 1998. "Von Börsenträumen, Wundermitteln und waghalsigen Versprechungen." *Frankfurter Allgemeine Zeitung*, July 8, 1998, 28.
Frantz, Simon. 2005. "2004 Approvals: The Demise of the Blockbuster?" *Nature Reviews Drug Discovery* 4: 93–94.
Franzosi, Roberto. 1998. "Narrative Analysis—or Why (and How) Sociologists Should Be Interested in Narrative." *Annual Review of Sociology* 24: 517–554.
Franzosi, Roberto. 2010. *Quantitative Narrative Analysis*. Thousand Oaks, CA: SAGE.
Fuhse, Jan A. 2015a. "Networks from Communication." *European Journal of Social Theory* 18: 39–59.
Fuhse, Jan A. 2015b. "Theorizing Social Networks: The Relational Sociology of and around Harrison White." *International Review of Sociology* 25: 15–44.
Fuhse, Jan A. 2021. *Social Networks of Meaning and Communication*. New York: Oxford University Press.
Fuhse, Jan A./Sophie Mützel (eds.). 2010. *Relationale Soziologie: Zur kulturellen Wende der Netzwerkforschung*. Wiesbaden: VS Verlag.
Fujimura, Joan H. 1996. *Crafting Science: A Sociohistory of the Quest for the Genetics of Cancer*. Cambridge, MA: Harvard University Press.
Gabriel, Yiannis. 2000. *Storytelling in Organizations: Facts, Fictions, and Fantasies*. Oxford: Oxford University Press.
Galambos, Louis/Jeffrey L. Sturchio. 1996. "The Pharmaceutical Industry in the Twentieth Century: A Reappraisal of the Sources of Innovation." *History and Technology* 13: 83–100.
Galambos, Louis/Jeffrey L. Sturchio. 1998. "Pharmaceutical Firms and the Transition to Biotechnology: A Study in Strategic Innovation." *Business History Review* 72: 250–278.
Galison, Peter. 1997. *Image and Logic: A Material Culture of Microphysics*. Chicago: University of Chicago Press.
Garavaglia, Christian/Franco Malerba/Luigi Orsenigo/Michele Pezzoni. 2012. "Technological Regimes and Demand Structure in the Evolution of the Pharmaceutical Industry." *Journal of Evolutionary Economics* 22: 677–709.
Garber, Ken. 2005. "New Apoptosis Drugs Face Clinical Test." *Nature Biotechnology* 23: 409–411.
Garber, Ken. 2013. "PARP Inhibitors Bounce Back." *Nature Reviews Drug Discovery* 12: 725–727.
Garber, Ken. 2014. "The Cancer Drug That Almost Wasn't." *Science* 345: 865–867.
Garcia-Parpet, Marie-France. 2010. "The Social Construction of a Perfect Market: The Strawberry Auction at Fontaines-en-Sologne," in Donald MacKenzie/Fabian

Muniesa/Lucia Siu (eds.), *Do Economists Make Markets? On the Performativitiy of Economics.* Princeton, NJ: Princeton University Press: 20–53.

Garud, Raghu/Joel Gehman/Antonio Paco Giuliani. 2014a. "Contextualizing Entrepreneurial Innovation: A Narrative Perspective." *Research Policy* 43: 1177–1188.

Garud, Raghu/Joel Gehman/Arun Kumaraswamy. 2011. "Complexity Arrangements for Sustained Innovation: Lessons from 3M Corporation." *Organization Studies* 32: 737–767.

Garud, Raghu/Cynthia Hardy/Steve Maguire. 2007. "Institutional Entrepreneurship as Embedded Agency: An Introduction to the Special Issue." *Organization Studies* 28: 957–969.

Garud, Raghu/Sanjay Jain/Arun Kumaraswamy. 2002. "Institutional Entrepreneurship in the Sponsorship of Common Technological Standards: The Case of Sun Microsystems and Java." *Academy of Management Journal* 45: 196–214.

Garud, Raghu/Peter Karnoe. 2001. "Path Creation as a Process of Mindful Deviation," in Raghu Garud/Peter Karnoe (eds.), *Path Dependence and Creation.* Mahwah, NJ: Erlbaum: 1–38.

Garud, Raghu/Michael A. Rappa. 1994. "A Sociocognitive Model of Technology Evolution—the Case of Cochlear Implants." *Organization Science* 5: 344–362.

Garud, Raghu/Henri A. Schildt/Theresa K. Lant. 2014b. "Entrepreneurial Storytelling, Future Expectations, and the Paradox of Legitimacy." *Organization Science* 25: 1479–1492.

Garud, Raghu/Barbara Simpson/Ann Langley/Haridimos Tsoukas. 2015. *The Emergence of Novelty in Organizations.* Oxford: Oxford University Press.

Garud, Raghu/Philipp Tuertscher/Andrew H. Van De Ven. 2013. "Perspectives on Innovation Processes." *Academy of Management Annals* 7: 775–819.

Geertz, Clifford. 2001. "The Bazaar Economy: Information and Search in Peasant Marketing," in Mark Granovetter/Richard Swedberg (eds.), *The Sociology of Economic Life.* Boulder, CO: Westview: 139–145.

Gibbs, Jackson B. 2000. "Mechanism-Based Target Identification and Drug Discovery in Cancer Research." *Science* 287: 1969–1973.

Gibson, David R. 2011a. "Avoiding Catastrophe: The Interactional Production of Possibility during the Cuban Missile Crisis." *American Journal of Sociology* 117: 361–419.

Gibson, David R. 2011b. "Speaking of the Future: Contentious Narration During the Cuban Missile Crisis." *Qualitative Sociology* 34: 503–522.

Gibson, David R. 2012. *Talk at the Brink: Deliberation and Decision During the Cuban Missile Crisis.* Princeton, NJ: Princeton University Press.

Gilbert, Denise. 1992. "Biotechnology: What Now?" *Nature Biotechnology* 10: 242–243.

Ginsburg, Geoffrey S./Jeanette J. McCarthy. 2001. "Personalized Medicine: Revolutionizing Drug Discovery and Patient Care." *Trends in Biotechnology* 19: 491–496.

Giorgi, Simona/Klaus Weber. 2015. "Marks of Distinction: Framing and Audience Appreciation in the Context of Investment Advice." *Administrative Science Quarterly* 60: 333–367.

Gittelman, Michelle/Bruce Kogut. 2003. "Does Good Science Lead to Valuable Knowledge? Biotechnology Firms and the Evolutionary Logic of Citation Patterns." *Management Science* 49: 366.
Gladwell, Malcolm. 1988. "New 'Wonder Drugs' Pose Patent Questions; Companies Turn to Courts to Resolve Who Owns Biotech Products." *The Washington Post*, August 5, 1988, A1.
Gladwell, Malcolm. 2010. "The Treatment: Why Is It So Difficult to Develop Drugs for Cancer?" *The New Yorker*: 68–77.
Glynn, Mary Ann/Chad Navis. 2013. "Categories, Identities, and Cultural Classification: Moving Beyond a Model of Categorical Constraint." *Journal of Management Studies* 50: 1124–1137.
Godart, Frédéric/Harrison C. White. 2010. "Switchings under Uncertainty: The Coming and Becoming of Meanings." *Poetics* 38: 567–586.
Goldstein, Jeffrey. 1999. "Emergence as a Construct: History and Issues." *Emergence* 1: 49–72.
Goldstone, Andrew/Ted Underwood. 2014. "The Quiet Transformations of Literary Studies: What Thirteen Thousand Scholars Could Tell Us." *New Literary History* 45: 359–384.
Grabher, Gernot. 1993. "The Weakness of Strong Ties: The Lock-ins of Regional Development in the Ruhr Area," in Gernot Grabher (ed.), *The Embedded Firm*. London: Routledge: 265–277.
Graham, Shawn/Ian Milligan/Scott Weingart. 2016. *Exploring Big Historical Data: The Historian's Macroscope*. London: Imperial College Press.
Granovetter, Mark. 1985. "Economic Action and Social Structure. The Problem of Embeddedness." *American Journal of Sociology* 91: 481–510.
Granovetter, Mark. 1995. *Getting a Job: A Study of Contacs and Careers*. Cambridge, MA: Harvard University Press.
Granqvist, Nina/Stine Grodal/Jennifer L. Woolley. 2013. "Hedging Your Bets: Explaining Executives' Market Labeling Strategies in Nanotechnology." *Organization Science* 24: 395–413.
Grant, David/Cynthia Hardy/Cliff Oswick/Linda L. Putnam (eds.). 2004. *The SAGE Handbook of Organizational Discourse*. London: SAGE.
Green, Daniel. 1994. "British Bio Seeks Pounds 93.6m to Fund Drugs Development." *Financial Times*, March 30, 1994.
Grimmer, Justin. 2010. "A Bayesian Hierarchical Topic Model for Political Texts: Measuring Expressed Agendas in Senate Press Releases." *Political Analysis* 18: 1–35.
Grimmer, Justin/Gary King. 2011. "General Purpose Computer-Assisted Clustering and Conceptualization." *Proceedings of the National Academy of Sciences of the United States of America* 108: 2643–2650.
Grimmer, Justin/Brandon M. Stewart. 2013. "Text as Data: The Promise and Pitfalls of Automatic Content Analysis Methods for Political Texts." *Political Analysis* 21: 267–297.

Grimmer, Justin/Margaret E. Roberts/Brandon M. Stewart. 2022. *Text as Data: A New Framework for Machine Learning and the Social Sciences*. Princeton, NJ: Princeton University Press.

Grodal, Stine/Aleksios Gotsopoulos/Fernando Suarez. 2015. "The Co-evolution of Technologies and Categories during Industry Emergence." *Academy of Management Review* 40: 423–445.

Grodal, Stine/Steven J. Kahl. 2017. "The Discursive Perspective of Market Categorization: Interaction, Power, and Context," in Rodolphe Durand/Nina Granqvist/Anna Tyllström (eds.), *From Categories to Categorization: Studies in Sociology, Organizations and Strategy at the Crossroads*. Bingley: Emerald Publishing 51: 151–184.

Grünberg, Jaan/Josef Pallas. 2013. "Beyond the News Desk—The Embeddedness of Business News." *Media, Culture & Society* 35: 216–233.

Gulati, Ranjay/Martin Gargiulo. 1999. "Where Do Interorganizational Networks Come From?" *American Journal of Sociology* 104: 1439–1493.

Gura, Trisha. 2002. "Therapeutic Antibodies: Magic Bullets Hit the Target." *Nature* 417: 584–586.

Haber, Daniel A./Nathanael S. Gray/Jose Baselga. 2011. "The Evolving War on Cancer." *Cell* 145: 19–24.

Hall, Carl T. 1996. "Good News on Biotech Breast Cancer Drug." *The San Francisco Chronicle*, March 1, 1996.

Hall, Carl T. 1998. "Swift OK From FDA on Drug to Treat Breast Cancer." *The San Francisco Chronicle*, September 26, 1998.

Hall, Stephen S. 1987. *Invisible Frontiers: The Race to Synthesize a Human Gene*. New York: Atlantic Monthly Press.

Han, Shin-Kap/Phyllis Moen. 1999. "Clocking Out: Temporal Patterning of Retirement." *American Journal of Sociology* 105: 191–236.

Hanahan, Douglas/Robert A. Weinberg. 2000. "The Hallmarks of Cancer." *Cell* 100: 57–70.

Hanahan, Douglas/Robert A. Weinberg. 2011. "Hallmarks of Cancer: The Next Generation." *Cell* 144: 646–674.

Hawkes, Nigel. 1994. "Treatment Curbs Spread of Cancer." *The Times*, March 19, 1994.

Hedgecoe, Adam. 2004. *The Politics of Personalized Medicine: Pharmacogenetics in the Clinic*. Cambridge: Cambridge University Press.

Hedgecoe, Adam/Paul Martin. 2003. "The Drugs Don't Work: Expectations and the Shaping of Pharmacogenetics." *Social Studies of Science* 33: 327–364.

Heintz, Bettina. 2004. "Emergenz und Reduktion: Neue Perspektiven auf das Mikro-Makro-Problem." *Kölner Zeitschrift für Soziologie und Sozialpsychologie* 56: 1–32.

Henderson, Rebecca/Luigi Orsenigo/Gary Pisano. 1999. "The Pharmaceutical Industry and the Revolution of Molecular Biology: Interactions among Scientific, Insitutional, and Organizational Change," in David Mowery/Richard Nelson (eds.), *The Sources of Industrial Leadership*. Cambridge: Cambridge University Press: 267–311.

Herrman, John/Kellen Browing. 2021. "Are We in the Metaverse Yet?" *The New York Times*, July 10, 2021.

Hodgson, John. 1990. "European Biopharmaceutical Culture." *Nature Biotechnology* 8: 720–723.
Hodgson, John. 1995. "Remodeling MMPIs." *Nature Biotechnology* 13: 554–557.
Holland, John H. 1998. *Emergence: From Chaos to Order*. Oxford: Oxford University Press.
Holmes, Douglas R. 2009. "Economy of Words." *Cultural Anthropology* 24: 381–419.
Holstein, James A./Jaber Gubrium. 2012. *Varieties of Narrative Analysis*. Los Angeles: SAGE.
Hope, Jenny. 1998. "Jab Could Prolong Lives of Patients, Say Doctors." *Daily Mail*, October 2, 1998.
Hsu, Greta/Stine Grodal. 2021. "The Double-edged Sword of Oppositional Category Positioning: A Study of the US E-cigarette Category, 2007–2017." *Administrative Science Quarterly* 66: 86–132.
Hsu, Greta/Peter W. Roberts/Anand Swaminathan. 2012. "Evaluative Schemas and the Mediating Role of Critics." *Organization Science* 23: 83–97.
Huber, Christoph H./Thomas Wölfel. 2004. "Immmunotherapy of Cancer: From Vision to Standard Clinical Practice." *Journal of Cancer Research and Clinical Oncology* 130: 367–374.
Huberman, Gur/Tomer Regev. 2001. "Contagious Speculation and a Cure for Cancer: A Nonevent That Made Stock Prices Soar." *Journal of Finance* 56: 387–396.
Hughes, Sally Smith. 2006. *Interview with Axel Ullrich, 1994 and 2003*. Berkeley: Regents of the University of California, http://bancroft.berkeley.edu/ROHO/projects/biosci/oh_list.html.
Hughes, Sally Smith. 2011. *Genentech: The Beginnings of Biotech*. Chicago: University of Chicago Press.
Hutter, Michael. 2011. "Infinite Surprises: On the Stabilization of Value in the Creative Industries," in Jens Beckert/Patrik Aspers (eds.), *The Worth of Goods*. Oxford: Oxford University Press: 201–221.
Hutter, Michael/David Stark. 2015. "Pragmatist Perspectives on Valuation: An Introduction," in Ariane Berthoin Antal/Michael Hutter/David Stark (eds.), *Moments of Valuation*. Oxford: Oxford University Press: 1–12.
IARC/WHO. 2020a. "Cancer Today," GLOBOCAN 2020: Estimated cancer incidence, mortality and prevalence worldwide in 2020, https://gco.iarc.fr/.
IARC/WHO. 2020b. "Cancer Tomorrow," GLOBOCAN 2020: Estimated cancer incidence, mortality and prevalence worldwide in 2020, https://gco.iarc.fr/tomorrow/en.
Investors Chronicle. 1994a. "Bearbull: The Greedy Stewards." *Investors Chronicle*, March 25, 1994.
Investors Chronicle. 1994b. "British Biotech—Patient Hope." *Investors Chronicle*, September 23, 1994.
Jackson, Matthew O. 2008. *Social and Economic Networks*. Princeton, NJ: Princeton University Press.
Jarzabkowski, Paula/John A. A. Sillince/Duncan Shaw. 2010. "Strategic Ambiguity as a Rhetorical Resource for Enabling Multiple Interests." *Human Relations* 63: 219–248.

Jockers, Matthew L. 2013. *Macroanalysis: Digital Methods and Literary History*. Urbana: University of Illinois Press.

Johnson, Steven. 2001. *Emergence: The Connected Lives of Ants, Brains, Cities, and Software*. New York: Scribner.

Jones, Candace/Massimo Maoret/Felipe G. Massa/Silviya Svejenova. 2012. "Rebels with a Cause: Formation, Contestation, and Expansion of the De Novo Category 'Modern Architecture,' 1870–1975." *Organization Science* 23: 1523–1545

Jong, Simcha. 2006. "How Organizational Structures in Science Shape Spin-off Firms: The Biochemistry Departments of Berkeley, Stanford, and UCSF and the Birth of the Biotech Industry." *Industrial and Corporate Change* 15: 251–283.

Jorgensen, Bud. 1992. "Biomira Hype May Require Grain of Salt." *The Globe and Mail*, January 29, 1992.

Jørgensen, Jan Trøst/Henrik Winther. 2009. "The New Era of Personalized Medicine: 10 Years Later." *Personalized Medicine* 6: 423–428.

Jurafsky, Dan/James Martin. 2009. *Speech and Natural Language Processing: An Introduction to Natural Language Processing, Computational Linguistics, and Speech Recognition*. 2nd. ed. Upper Saddle River, NJ: Pearson Prentice-Hall.

Kamb, Alexander/Susan Wee/Christoph Lengauer. 2007. "Why Is Cancer Drug Discovery So Difficult?" *Nature Reviews Drug Discovery* 6: 115–120.

Kaplan, Sarah/Fiona Murray. 2010. "Entrepreneurship and the Construction of Value in Biotechnology." *Research in the Sociology of Organizations* 29: 107–147.

Kaplan, Sarah/Wanda J. Orlikowski. 2013. "Temporal Work in Strategy Making." *Organization Science* 24: 965–995.

Kaplan, Sarah/Mary Tripsas. 2008. "Thinking about Technology: Applying a Cognitive Lens to Technical Change." *Research Policy* 37: 790–805.

Kaplan, Sarah/Keyvan Vakili. 2014. "The Double-Edged Sword of Recombination in Breakthrough Innovation." *Strategic Management Journal* 36: 1435–1457.

Karell, Daniel/Michael Freedman. 2019. "Rhetorics of Radicalism." *American Sociological Review* 84: 726–753.

Karell, Daniel/Michael Freedman. 2020. "Sociocultural Mechanisms of Conflict: Combining Topic and Stochastic Actor-Oriented Models in an Analysis of Afghanistan, 1979–2001." *Poetics* 78: 101403–101403.

Kaufman, Jason. 2004. "Endogenous Explanation in the Sociology of Culture." *Annual Review of Sociology* 30: 335–357.

Keating, Peter/Alberto Cambrosio. 2012. *Cancer on Trial: Oncology as a New Style of Practice*. Chicago: University of Chicago Press.

Keller, Reiner. 2013. *Doing Discourse Research: An Introduction for Social Scientists*. London; Thousand Oaks, CA: SAGE.

Kennedy, Mark T. 2005. "Behind the One-Way Mirror: Refraction in the Construction of Product Market Categories." *Poetics* 33: 201–226.

Kennedy, Mark T. 2008. "Getting Counted: Markets, Media, and Reality." *American Sociological Review* 73: 270–295.

Kennedy, Mark T./Peer C. Fiss. 2013. "An Ontological Turn in Categories Research: From Standards of Legitimacy to Evidence of Actuality." *Journal of Management Studies* 50: 1138–1154.

Kenney, Martin. 1986a. *Biotechnology: The University-Industrial Complex*. New Haven, CT: Yale University Press.

Kenney, Martin. 1986b. "Schumpeterian Innovation and Entrepreneurs in Capitalism: A Case Study of the US Biotechnology Industry." *Research Policy* 15: 21–31.

Kenney, Martin/Richard Florida. 2000. "Venture Capital in Silicon Valley: Fueling New Firm Formation," in Martin Kenney (ed.), *Understanding Silicon Valley: The Anatomy of an Entrepreneurial Region*. Stanford, CA: Stanford University Press: 98–123.

Khaire, Mukti/R. Daniel Wadhwani. 2010. "Changing Landscapes: The Construction of Meaning and Value in a New Market Category—Modern Indian Art." *Academy of Management Journal* 53: 1281–1304.

Kho, Patricia/Wei Chua/Melissa M. Morre/Stephen J. Clarke. 2010. "Is It Prime Time for Personalized Medicine in Cancer Treatment?" *Personalized Medicine* 7: 387–397.

Klamer, Arjo/Deirdre N. McCloskey/Robert M. Solow (eds.). 1988. *The Consequences of Economic Rhetoric*. Cambridge: Cambridge University Press.

Klausner, Arthur. 1986. "Taking Aim at Cancer with Monoclonal Antibodies." *Nature Biotechnology* 4: 185–194.

Klausner, Arthur. 1988. "Biotech Analysts' Predictions for '88." *Nature Biotechnology* 6: 32–36.

Klawiter, Maren. 2008. *The Biopolitics of Breast Cancer: Changing Cultures of Disease and Activism*. Minneapolis: University of Minnesota Press.

Knorr Cetina, Karin. 2006. "The Market." *Theory, Culture & Society* 23: 551–556.

Knorr Cetina, Karin/Urs Bruegger. 2002. "Global Microstructures: The Virtual Societies of Financial Markets." *American Journal of Sociology* 107: 905–950.

Knorr Cetina, Karin/Aaron V. Cicourel (eds.). 1981. *Advances in Social Theory and Methodology: Toward an Integration of Micro- and Macro-Sociologies*. Boston: Routledge & Kegan Paul.

Knorr Cetina, Karin/Alex Preda. 2005. "Introduction," in Karin Knorr Cetina/Alex Preda (eds.), *The Sociology of Financial Markets*. Oxford: Oxford University Press: 1–14.

Knowles, Jonathan/Gianni Gromo. 2003. "Target Selection in Drug Discovery." *Nature Reviews Drug Discovery* 2: 63–69.

Kolata, Gina. 1998. "Hope in the Lab: A Special Report. A Cautious Awe Greets Drugs That Eradicate Tumors in Mice." *The New York Times*, May 3, 1998, 1.

Kolata, Gina. 2012. "Study Reshaping Ways of Treating Breast Cancer." *The New York Times*, September 24, 2012, A1.

Kolker, Emily S. 2004. "Framing as a Cultural Resource in Health Social Movements: Funding Activism and the Breast Cancer Movement in the US 1990–1993." *Sociology of Health & Illness* 26: 820–844.

Krippendorff, Klaus. 2012. *Content Analysis: An Introduction to Its Methodology*. Thousand Oaks, CA: SAGE.

Kuckartz, Udo. 2004. "Computerunterstützte Analyse qualitativer Daten," in Andreas Diekmann (ed.), *Methoden der Sozialforschung. Sonderheft 44/2004 der KZfSS*. Wiesbaden: VS Verlag für Sozialwissenschaften: 453–478.

Kuckartz, Udo. 2007. *Einführung in die computergestützte Analyse qualitativer Daten*. 2nd. ed. Wiesbaden: VS Verlag für Sozialwissenschaften.

Kushner, Rose. 1975. *Breast Cancer: A Personal History and an Investigative Report*. New York: Harcourt Brace Jovanovich.

Lacroix, Marc. 2011. *A Concise History of Breast Cancer*. New York: Nova Science Publishers.

Lamont, Michèle. 2000. "Meaning-Making in Cultural Sociology: Broadening Our Agenda." *Contemporary Sociology* 29: 602–607.

Lamont, Michèle. 2012. "Toward a Comparative Sociology of Valuation and Evaluation." *Annual Review of Sociology* 38: 201–221.

Lamont, Michèle/Virag Molnar. 2002. "The Study of Boundaries in the Social Sciences." *Annual Review of Sociology* 28: 167–195.

Lane, David A./Robert R. Maxfield. 2005. "Ontological Uncertainty and Innovation." *Journal of Evolutionary Economics* 15: 3–50.

Langreth, Robert/Michael Waldholz. 1999. "New Era of Personalized Medicine Targeting Drugs for Each Unique Genetic Profile." *Oncologist* 4: 426–427.

Lanthier, Jennifer. 1996. "Tests Positive for Biomira Breast Cancer Treatment." *The Financial Post*, November 9, 1996.

Lanthier, Jennifer. 1997. "Biomira Confirms Partnership with Chiron." *The Financial Post*, May 6, 1997.

Latour, Bruno. 1986. "The Powers of Association," in John Law (ed.), *Power, Action and Belief*. London, Boston, and Henley: Routledge & Kegan Paul: 264–280.

Latour, Bruno. 1988. *The Pasteurization of France*. Cambridge, MA: Harvard University Press.

Latour, Bruno. 1996. *Aramis, or, The Love of Technology*. Cambridge, MA: Harvard University Press.

Latour, Bruno. 1999. "On Recalling ANT," in John Law/John Hassard (eds.), *Actor Network Theory and After*. Oxford: Blackwell Publishers: 15–25.

Latour, Bruno. 2005. *Reassembling the Social: An Introduction to Actor-Network Theory*. Oxford: Oxford University Press.

Latour, Bruno/Pablo Jensen/Tommaso Venturini/Sébastian Grauwin/Dominique Boullier. 2012. "The Whole Is Always Smaller than Its Parts: A Digital Test of Gabriel Tarde's Monads." *British Journal of Sociology* 63: 590–615.

Law, John/John Hassard (eds.). 1999. *Actor Network Theory and After*. Oxford: Blackwell.

Lawrence, Stacy. 2007a. "Billion Dollar Babies—Biotech Drugs as Blockbusters." *Nature Biotechnology* 25: 380–382.

Lawrence, Stacy. 2007b. "Pipelines Turn to Biotech." *Nature Biotechnology* 25: 1342–1342.

Lazer, David M. J./Alex Pentland/Duncan J. Watts/Sinan Aral/Susan Athey/Noshir Contractor/Deen Freelon/Sandra Gonzalez-Bailon/Gary King/Helen Margetts/Alondra Nelson/Matthew J. Salganik/Markus Strohmaier/Alessandro Vespignani/Claudia Wagner. 2020. "Computational Social Science: Obstacles and Opportunities." *Science* 369: 1060–1062.

Lazer, David/Alex Pentland/Lada Adamic/Sinan Aral/Albert-Laszlo Barabasi/Devon Brewer/Nicholas Christakis/Noshir Contractor/James H. Fowler/Myron Gutmann/Tony Jebara/Gary King/Michael Macy/Deb Roy/Marshall Van Alstyne. 2009. "Computational Social Science." *Science* 323: 721–723.

Ledford, Heidi. 2012. "Drug Candidates Derailed in Case of Mistaken Identity." *Nature* 483: 519.

Leifer, Eric. 1985. "Markets as Mechanisms: Using a Role Structure." *Social Forces* 64: 442–472.

Leifer, Eric. 1988. "Interaction Preludes to Role Setting: Exploratory Local Action." *American Sociological Review* 53: 865–878.

Leifer, Eric/Valli Rajah. 2000. "Getting Observations: Strategic Ambiguities in Social Interaction." *Soziale Systeme* 6: 251–267.

Leifer, Eric/Harrison C. White. 1987. "A Structural Approach to Markets," in Mark Mizruchi/Michael Schwartz (eds.), *Intercorporate Relations*. Cambridge: Cambridge University Press: 85–108.

Lengauer, Christoph/Luis A. Diaz/Saurabh Saha. 2005. "Cancer Drug Discovery through Collaboration." *Nature Reviews Drug Discovery* 4: 375–380.

Lepinay, Vincent-Antonin. 2011. *Codes of Finance: Engineering Derivatives in a Global Bank*. Princeton, NJ: Princeton University Press.

Lerner, Barron H. 2001. *The Breast Cancer Wars: Hope, Fear, and the Pursuit of a Cure in Twentieth-Century America*. New York: Oxford University Press.

Lerner, Barron H. 2002. "Breast Cancer Activism: Past Lessons, Future Directions." *Nature Reviews Cancer* 2: 225–230.

Light, Ryan/jimi adams. 2016. "Knowledge in Motion: The Evolution of HIV/AIDS Research." *Scientometrics* 107: 1227–1248.

Lizardo, Omar. 2006. "How Cultural Tastes Shape Personal Networks." *American Sociological Review* 71: 778–807.

Loewenstein, Jeffrey/William Ocasio/Candace Jones. 2012. "Vocabularies and Vocabulary Structure: A New Approach Linking Categories, Practices, and Institutions." *Academy of Management Annals* 6: 41–86.

Lord, Christopher J./Alan Ashworth. 2016. "BRCAness Revisited." *Nature Reviews Cancer* 16: 110–120.

Lorraine, Francois/Harrison C. White. 1971. "Structural Equivalence of Individuals in Social Networks." *Journal of Mathematical Sociology* 1: 49–80.

Lounsbury, Michael/Mary Ann Glynn. 2001. "Cultural Entrepreneurship: Stories, Legitimacy, and the Acquisition of Resources." *Strategic Management Journal* 22: 545–564.

Lowy, Douglas R./Francis S. Collins. 2016. "Aiming High—Changing the Trajectory for Cancer." *New England Journal of Medicine*, April 2016.

Löwy, Ilana. 2011. *A Woman's Disease: The History of Cervical Cancer*. Oxford: Oxford University Press.

Lumsden, Quentin. 1994. "Shares: Risky Flights in Blue Skies." *The Independent*, October 23, 1994, 12.

MacKenzie, Donald. 2006. *An Engine, Not a Camera: How Financial Models Shape Markets*. Cambridge, MA: MIT Press.

MacKenzie, Donald/Yuval Millo. 2003. "Constructing a Market, Performing Theory: The Historical Sociology of a Financial Derivatives Exchange." *American Journal of Sociology* 109: 107–145.

MacKenzie, Donald/Fabian Muniesa/Lucia Siu (eds.). 2007. *Do Economists Make Markets? On the Performativity of Economics*. Princeton, NJ: Princeton University Press.

Maesse, Jens. 2013. "Spectral Performativity: How Economic Expert Discourse Constructs Economic Worlds." *Economic Sociology: The European Electronic Newsletter* 14: 25–31.

Malerba, Franco/Luigi Orsenigo. 2002. "Innovation and Market Structure in the Dynamics of the Pharmaceutical Industry and Biotechnology: Towards a History-Friendly Model." *Industrial and Corporate Change* 11: 667–703.

Manley, John. 1998. "Chiroscience Provides Contrast in Strategy with British Biotech." *Extel Examiner*, April 29, 1998.

Marks, Lara V. 2015. *The Lock and Key of Medicine: Monoclonal Antibodies and the Transformation of Healthcare*. London, New Haven, CT: Yale University Press.

Marks, Paul A. 2014. *On the Cancer Frontier: One Man, One Disease, and a Medical Revolution*. With James Sterngold. New York: PublicAffairs.

Marsh, Peter. 1989. "Approval of Biotech Drug in Germany Gives Fillip to Industry." *Financial Times*, January 4, 1990.

Marshall, Lee. 2000. *The Cancer Outlook 2000*. London: Business Insights.

Martin, John Levi. 2000. "What Do Animals Do All Day? The Division of Labor, Class Bodies, and Totemic Thinking in the Popular Imagination." *Poetics* 27: 195–231.

Martin, Paul. 2015. "Commercialising Neurofutures: Promissory Economies, Value Creation and the Making of a New Industry." *BioSocieties* 10: 422–443.

Massod, Ehsan. 1998. "UK Biotech Flagship Meets Stormy Waters." *Nature* 392: 746.

Maurer, Indre/Mark Ebers. 2006. "Dynamics of Social Capital and Their Performance Implications: Lessons from Biotechnology Start-Ups." *Administrative Science Quarterly*: 262–292.

Maxmen, Amy. 2012a. "Cancer Research: Open Ambition." *Nature* 488: 148–150.

Maxmen, Amy. 2012b. "The Hard Facts." *Nature* 485: S50–S51.

Mayring, Philipp. 2015. "Qualitative Content Analysis: Theoretical Background and Procedures," in Angelika Bikner-Ahsbahs/Christine Knipping/Norma Presmeg (eds.), *Approaches to Qualitative Research in Mathematics Education*. Dordrecht: Springer: 365–380.

McCloskey, Deirdre N. 1998. *The Rhetoric of Economics*. Madison: University of Wisconsin Press.

McCloskey, Deirdre N. 1990a. *If You're So Smart*. Chicago: University of Chicago Press.
McCloskey, Deirdre N. 1990b. "Storytelling in Economics," in Christopher Nash (ed.), *Narrative in Culture: The Uses of Storytelling in the Sciences, Philosophy, and Literature*. London: Routledge: 5–22.
McCloskey, Deirdre N. 1995. "Metaphors Economists Live By." *Social Research* 62: 215–237.
McCloskey, Deirdre N./Arjo Klamer. 1995. "One Quarter of GDP Is Persuasion." *American Economic Review* 85: 191–195.
Mccrone, Angus. 1994. "Biotech on Trial for Its Future." *Evening Standard*, November 10, 1994, 38.
McFarland, Daniel A./Kevin Lewis/Amir Goldberg. 2016. "Sociology in the Era of Big Data: The Ascent of Forensic Social Science." *American Sociologist* 47: 12–35.
McFarland, Daniel A./Daniel Ramage/Jason Chuang/Jeffrey Heer/Christopher D. Manning/Daniel Jurafsky. 2013. "Differentiating Language Usage through Topic Models." *Poetics* 41: 607–625.
McKelvey, Maureen. 1996. *Evolutionary Innovations: The Business of Biotechnology*. Oxford: Oxford University Press.
McLaren-Hankin, Yvonne. 2008. "'We Expect to Report on Significant Progress in Our Product Pipeline in the Coming Year': Hedging Forward-Looking Statements in Corporate Press Releases." *Discourse Studies* 10: 635–654.
McLean, Paul D. 2007. *The Art of the Network. Strategic Interaction and Patronage in Renaissance Florence*. Durham, NC: Duke University Press.
Meeks, Elijah/Scott B. Weingart. 2013. "The Digital Humanities Contribution to Topic Modeling." *Journal of Digital Humanities* 1, http://journalofdigitalhumanities.org/2-1/dh-contribution-to-topic-modeling/.
Mellman, Ira/George Coukos/Glenn Dranoff. 2011. "Cancer Immunotherapy Comes of Age." *Nature* 480: 480–489.
Melnikova, Irena/James Golden. 2004. "Apoptosis-Targeting Therapies." *Nature Reviews Drug Discovery* 3: 905–906.
Mertens, Gilbert. 2004. *Targeted Cancer Therapies*. London: Business Insights.
Mertens, Gilbert. 2005. *Beyond the Blockbuster Drug: Strategies for Nichebuster Drugs, Targeted Therapies and Personalized Medicine*. London: Business Insights.
Metzler, Ingrid. 2010. "Biomarkers and Their Consequences for the Biomedical Profession: A Social Science Perspective." *Personalized Medicine* 7: 407–420.
Miles, Matthew B./A. Michael Huberman. 1994. *Qualitative Data Analysis: An Expanded Sourcebook*. 2nd ed. Thousand Oaks, CA: SAGE.
Miles, Matthew B./A. Michael Huberman/Johnny Saldaña. 2013. *Qualitative Data Analysis: A Methods Sourcebook*. 3rd ed. Thousand Oaks, CA: SAGE.
Mische, Ann. 2003. "Cross-Talk in Movements: Reconceiving the Culture-Network Link," in Mario Diani/Doug McAdam (eds.), *Social Movements and Networks*. Oxford: Oxford University Press: 258–280.
Mische, Ann. 2008. *Partisan Publics: Communication and Contention across Brazilian Youth Activist Networks*. Princeton, NJ: Princeton University Press.

Mische, Ann. 2009. "Projects and Possibilities: Researching Futures in Action." *Sociological Forum* 24: 694–704.
Mische, Ann. 2011. "Relational Sociology, Culture, and Agency," in John Scott/Peter J. Carrington (eds.), *The SAGE Handbook of Social Network Analysis*. London: SAGE: 80–97.
Mische, Ann/Harrison C. White. 1998. "Between Conversation and Situation: Public Switching Dynamics across Network Domains." *Social Research* 65: 695–724.
Mitchinson, W. 2005. "Breast Cancer in Canada: Medical Response and Attitudes, 1900–1950." *Histoire Sociale—Social History* 38: 399–432.
Mizruchi, Mark/Joseph Galaskiewicz. 1993. "Networks of Interorganizational Relations." *Sociological Methods & Research* 22: 46–70.
Mohr, John W./Christopher A. Bail/Margaret Frye/Jennifer C. Lena/Omar Lizardo/Terence E. McDonnell/Ann Mische/Iddo Tavory/Frederick F. Wherry. 2020. *Measuring Culture*. New York: Columbia University Press.
Mohr, John W. 1994. "Soldiers, Mothers, Tramps and Others: Discourse Roles in the 1907 New York City Charity Directory." *Poetics* 22: 327–357.
Mohr, John W. 1998. "Measuring Meaning Structures." *Annual Review of Sociology* 24: 345–370.
Mohr, John W./Petko Bogdanov. 2013. "Introduction—Topic Models: What They Are and Why They Matter." *Poetics* 41: 545–569.
Mohr, John W./Vincent Duquenne. 1997. "The Duality of Culture and Practice: Poverty Relief in New York City, 1888–1917." *Theory and Society* 26: 305–356.
Mohr, John W./Helene Lee. 2000. "From Affirmative Action to Outreach: Discourse Shifts at the University of California." *Poetics* 28: 47–71.
Mohr, John W./Craig Rawlings. 2015. *Formal Methods of Cultural Analysis. International Encyclopedia of the Social & Behavioral Sciences*. James D. Wright. Oxford: Elsevier, 5: 357–367.
Mohr, John W./Robin Wagner-Pacifici/Ronald L. Breiger. 2015. "Toward a Computational Hermeneutics." *Big Data & Society*, July–December 2015: 1–8.
Mohr, John W./Robin Wagner-Pacifici/Ronald L. Breiger/Petko Bogdanov. 2013. "Graphing the Grammar of Motives in National Security Strategies: Cultural Interpretation, Automated Text Analysis, and the Drama of Global Politics." *Poetics* 41: 670–700.
Moody, James. 2004. "The Structure of a Social Science Collaboration Network: Disciplinary Cohesion from 1963 to 1999." *American Sociological Review* 69: 213.
Moody, James/Ryan Light. 2006. "A View from Above: The Evolving Sociological Landscape." *American Sociologist* 37: 67–86.
Moody, James/Douglas R. White. 2003. "Structural Cohesion and Embeddedness: A Hierarchical Concept of Social Groups." *American Sociological Review* 68: 103–127.
Moretti, Franco. 2013. *Distant Reading*. London: Verso.
Morrison, Michael/Lucas Cornips. 2012. "Exploring the Role of Dedicated Online Biotechnology News Providers in the Innovation Economy." *Science, Technology & Human Values* 37: 262–285.

Moy, Beverly/Peter Kirkpatrick/Santwana Kar/Paul Goss. 2007. "Lapatinib." *Nature Reviews Drug Discovery* 6: 431–432.
Muhsin, Mohamed/Joanne Graham/Peter Kirkpatrick. 2004. "Bevacizumab." *Nature Reviews Drug Discovery* 3: 995–996.
Mukherjee, Siddartha. 2010. *The Emperor of All Maladies*. New York: Scribner.
Muniesa, Fabian/Yuval Millo/Michel Callon. 2007. "An Introduction to Market Devices." *Sociological Review* 55: 1–12.
Mutch, Alistair/Rick Delbridge/Marc Ventresca. 2006. "Situating Organizational Action: The Relational Sociology of Organizations." *Organization* 13: 607–625.
Mützel, Sophie. 2002. *Making Meaning of the Move of the German Capital: Networks, Logics, and the Emergence of Capital City Journalism*. Ann Arbor, MI: UMI.
Mützel, Sophie. 2009. "Networks as Culturally Constituted Processes: A Comparison of Relational Sociology and Actor-network Theory." *Current Sociology* 57: 871–887.
Mützel, Sophie. 2010. "Koordinierung von Märkten durch narrativen Wettbewerb." *Kölner Zeitschrift für Soziologie und Sozialpsychologie, Sonderheft* 49: 87–106.
Mützel, Sophie. 2015. "Structures of the Tasted: Restaurant Reviews in Berlin Between 1995 and 2012," in Ariane Berthoin Antal/Michael Hutter/David Stark (eds.), *Moments of Valuation: Exploring Sites of Dissonance*. Oxford: Oxford University Press: 147–167.
Mützel, Sophie. 2021. "Unlocking the Payment Experience: Future Imaginaries in the Case of Digital Payments." *New Media & Society* 23: 284–301.
Mützel, Sophie/Ronald L. Breiger. 2020. "Duality beyond Persons and Groups: Culture and Affiliation," in Ryan Light/James Moody (eds.), *Oxford Handbook of Social Networks*. Oxford: Oxford University Press: 392–413.
Mützel, Sophie/Lisa Kressin. 2021. "From Simmel to Relational Sociology," in Seth Abrutyn/Omar Lizardo (eds.), *Handbook of Classical Sociological Theory*. Cham, Switzerland: Springer: 217–238.
Nathan, David G. 2007. *The Cancer Treatment Revolution: How Smart Drugs and Other New Therapies Are Renewing Our Hope and Changing the Face of Medicine*. Hoboken, NJ: John Wiley & Sons.
Nature Biotechnology. 1988. "Biotechnology Products in the Pipeline." *Nature Biotechnology* 6: 1004–1007.
Navis, Chad/Mary Ann Glynn. 2010. "How New Market Categories Emerge: Temporal Dynamics of Legitimacy, Identity, and Entrepreneurship in Satellite Radio, 1990–2005." *Administrative Science Quarterly* 55: 439–471.
Negro, Giacomo/Özgecan Kocak/Greta Hsu. 2010. "Research on Categories in the Sociology of Organizations," in Greta Hsu/Özgecan Kocak/Giacomo Negro (eds.), *Categories in Markets: Origins and Evolution*. Bingley, UK: Emerald 31: 3–35.
Nelson, Laura K. 2020. "Computational Grounded Theory: A Methodological Framework." *Sociological Methods & Research* 49: 3–42.
Nelson, Laura K. 2021. "Cycles of Conflict, a Century of Continuity: The Impact of Persistent Place-Based Political Logics on Social Movement Strategy." *American Journal of Sociology* 127: 1–59.

Nelson, Richard R. 1993. *National Innovation Systems: A Comparative Study*. Oxford: Oxford University Press.
Newell, Sue/Anna Goussevskaia/Jacky Swan/Mike Bresnen/Ademola Obembe. 2008. "Interdependencies in Complex Project Ecologies: The Case of Biomedical Innovation." *Long Range Planning* 41: 33–54.
Nigam, Amit/William Ocasio. 2010. "Event Attention, Environmental Sensemaking, and Change in Institutional Logics: An Inductive Analysis of the Effects of Public Attention to Clinton's Health Care Reform Initiative." *Organization Science* 21: 823–841.
Nik-Zainal, Serena/Helen Davies/Johan Staaf/Manasa Ramakrishna/Dominik Glodzik/Xueqing Zou/Inigo Martincorena/Ludmil B. Alexandrov/Sancha Martin/David C. Wedge/Peter Van Loo/Young Seok Ju/Marcel Smid/Arie B. Brinkman/Sandro Morganella/Miriam R. Aure/Ole Christian Lingjærde/Anita Langerød/Markus Ringnér/Sung-Min Ahn/Sandrine Boyault/Jane E. Brock/Annegien Broeks/Adam Butler/Christine Desmedt/Luc Dirix/Serge Dronov/Aquila Fatima/John A. Foekens/Moritz Gerstung/Gerrit K. J. Hooijer/Se Jin Jang/David R. Jones/Hyung-Yong Kim/Tari A. King/Savitri Krishnamurthy/Hee Jin Lee/Jeong-Yeon Lee/Yilong Li/Stuart McLaren/Andrew Menzies/Ville Mustonen/Sarah O'Meara/Iris Pauporté/Xavier Pivot/Colin A. Purdie/Keiran Raine/Kamna Ramakrishnan/F. Germán Rodríguez-González/Gilles Romieu/Anieta M. Sieuwerts/Peter T. Simpson/Rebecca Shepherd/Lucy Stebbings/Olafur A. Stefansson/Jon Teague/Stefania Tommasi/Isabelle Treilleux/Gert G. Van den Eynden/Peter Vermeulen/Anne Vincent-Salomon/Lucy Yates/Carlos Caldas/Laura van't Veer/Andrew Tutt/Stian Knappskog/Benita Kiat Tee Tan/Jos Jonkers/Åke Borg/Naoto T. Ueno/Christos Sotiriou/Alain Viari/P. Andrew Futreal/Peter J. Campbell/Paul N. Span/Steven Van Laere/Sunil R. Lakhani/Jorunn E. Eyfjord/Alastair M. Thompson/Ewan Birney/Hendrik G. Stunnenberg/Marc J. van de Vijver/John W. M. Martens/Anne-Lise Børresen-Dale/Andrea L. Richardson/Gu Kong/Gilles Thomas/Michael R. Stratton. 2016. "Landscape of Somatic Mutations in 560 Breast Cancer Whole-Genome Sequences." *Nature* 534: 47–54.
Ocana, Alberto/Atanasio Pandiella/Lilian L. Siu/Ian F. Tannock. 2011. "Preclinical Development of Molecular-Targeted Agents for Cancer." *Nature Review of Clinical Oncology* 8: 200–209.
Ocasio, William/John Joseph. 2005. "Cultural Adaptation and Institutional Change: The Evolution of Vocabularies of Corporate Governance, 1972–2003." *Poetics* 33: 163–178.
Oliff, Allen/Jackson B. Gibbs/Frank McCormick. 1996. "New Molecular Targets for Cancer Therapy." *Scientific American*: 144–149.
Olson, James Stuart. 2002. *Bathsheba's Breast: Women, Cancer & History*. Baltimore: Johns Hopkins University Press.
Orr, Julian. 1996. *Talking about Machines: An Ethnography of a Modern Job*. Ithaca, NY: Cornell University Press.
Owen-Smith, Jason/Massimo Riccaboni/Fabio Pammolli/Walter W. Powell. 2002. "A Comparison of U.S. and European University-Industry Relations in the Life Sciences." *Management Science* 48: 24–43.

Owen-Smith, Jason/Walter W. Powell. 2004. "Knowledge Networks as Channels and Conduits: The Effects of Spillovers in the Boston Biotechnology Community." *Organization Science* 15: 5.

Pachucki, Mark A./Ronald L. Breiger. 2010. "Cultural Holes: Beyond Relationality in Social Networks and Culture." *Annual Review of Sociology* 36: 205–224.

Padgett, John F. 2012. "Transposition and Refunctionality: The Birth of Partnership Systems in Renaissance Florence," in John F. Padgett/Walter W. Powell (eds.), *The Emergence of Organizations and Markets*. Princeton, NJ: Princeton University Press: 168–207.

Padgett, John F. 2018. "Faulkner's Assembly of Memories into History: Narrative Networks in Multiple Times." *American Journal of Sociology* 124: 406–478.

Padgett, John F./Walter W. Powell. 2012a. *The Emergence of Organizations and Markets*. Princeton, NJ: Princeton University Press.

Padgett, John F./Walter W. Powell. 2012b. "The Problem of Emergence," in John F. Padgett/Walter W. Powell (eds.), *The Emergence of Organizations and Markets*. Princeton, NJ: Princeton University Press: 1–29.

Pammolli, Fabio/Laura Magazzini/Massimo Riccaboni. 2011. "The Productivity Crisis in Pharmaceutical R&D." *Nature Reviews Drug Discovery* 10: 428–438.

Penan, Hervé. 1996. "R & D Strategy in a Techno-Economic network: Alzheimer's Disease Therapeutic Strategies." *Research Policy* 25: 337–358.

Pentland, Brian T. 1999. "Building Process Theory with Narrative: From Description to Explanation." *Academy of Management Review* 24: 711–724.

Pharmaceutical Business News. 1989. "Canadian Cancer Vaccine Research Seeks for Funding." *Pharmaceutical Business News*, January 6, 1989.

Pharmaceutical Business News. 1991. "Genentech to Begin Trials on Cancer Therapy in the US." *Pharmaceutical Business News*, April 12, 1991.

Pharmaceutical Business News. 1995. "Chiroscience Group Reveals Promising New Product Pipeline." *Pharmaceutical Business News*, March 1, 1995.

Phillips, Nelson. 1995. "Telling Organizational Tales: On the Role of Narrative Fiction in the Study of Organizations." *Organization Studies* 16: 625–649.

Pickstone, John V. 2007. "Contested Cumulations: Configurations of Cancer Treatments through the Twentieth Century." *Bulletin of the History of Medicine* 81: 164–196.

Pieters, Toine. 1998. "Marketing Medicines through Randomised Controlled Trials: The Case of Interferon." *BMJ: British Medical Journal* 317: 1231.

Pieters, Toine. 2005. *Interferon: The Science and Selling of a Miracle Drug*. London; New York: Routledge.

Pisano, Gary P. 1991. "The Governance of Innovation: Vertical Integration and Collaborative Arrangements in the Biotechnology." *Research Policy* 20: 237–249.

Pisano, Gary P. 2006. *Science Business: The Promise, the Reality, and the Future of Biotech*. Boston: Harvard Business School Press.

Podolny, Joel M. 1993. "A Status-Based Model of Market Competition." *American Journal of Sociology* 98: 829–872.

Podolny, Joel M. 2001. "Networks as the Pipes and Prisms of the Market." *American Journal of Sociology* 107: 33–60.
Podolny, Joel M. 2005. *Status Signals: A Sociological Study of Market Competition*. Princeton, NJ: Princeton University Press.
Podolny, Joel M./Karen Page. 1998. "Network Forms of Organization." *Annual Review of Sociology* 24: 57–76.
Polletta, Francesca. 2009. *It Was Like a Fever: Storytelling in Protest and Politics*. Chicago: University of Chicago Press.
Pollock, Neil/Robin Williams. 2010. "The Business of Expectations: How Promissory Organizations Shape Technology and Innovation." *Social Studies of Science* 40: 525–548.
Pollock, Neil/Robin Williams. 2016. *How Industry Analysts Shape the Digital Future*. Oxford: Oxford University Press.
Pontikes, Elizabeth G. 2012. "Two Sides of the Same Coin: How Ambiguous Classification Affects Multiple Audiences' Evaluations." *Administrative Science Quarterly* 57: 81–118.
Powell, Walter W. 1990. "Neither Market nor Hierarchy: Network Forms of Organization." *Research in Organizational Behavior* 12: 295–336.
Powell, Walter W. 1996. "Inter-organizational Collaboration in the Biotechnology Industry." *Journal of Institutional and Theoretical Economics* 152: 197–215.
Powell, Walter W. 1998. "Learning from Collaboration: Knowledge and Networks in the Biotechnology and Pharmaceutical Industries." *California Management Review* 40: 228–240.
Powell, Walter W./Peter Brantley. 1992. "Competitive Cooperation in Biotechnology: Learning through Networks?," in Nitin Nohria/Robert Eccles (eds.), *Networks and Organizations: Structure, Form, and Action*. Cambridge, MA: Harvard Business School Press: 366–394.
Powell, Walter W./Peter Brantley. 1996. "Magic Bullets and Patent Wars: New Product Development and the Evolution of the Biotechnology Industry," in Toshihiro Nishiguchi (ed.), *Managing Product Development*. Oxford: Oxford University Press: 233–260.
Powell, Walter W./Kennth W. Koput/Laurel Smith-Doerr. 1996. "Interorganizational Collaboration and the Locus of Innovation: Networks of Learning in Biotechnology." *Administrative Science Quarterly* 41: 116–145.
Powell, Walter W./Kennth W. Koput/Laurel Smith-Doerr/Jason Owen-Smith. 1999. "Network Position and Firm Performance: Organizational Returns to Collaboration in the Biotechnology Industry," in Steven Andrews/David Knoke (eds.), *Networks in and around Organizations*. Greenwich, CT: JAI Press.
Powell, Walter W./Jason Owen-Smith. 1998. "Universities and the Market for Intellectual Property in the Life Sciences." *Journal of Policy Analysis and Management* 17: 253–277.
Powell, Walter W./Kelley Packalen/Kjersten Bunker Whittington. 2012. "Organizational and Institutional Genesis: The Emergence of High-tech Clusters in the Life

Sciences," in John F. Padgett/Walter W. Powell (eds.), *The Emergence of Organizations and Markets*. Princeton, NJ: Princeton University Press: 434–465.

Powell, Walter W./Kurt W. Sandholtz. 2012a. "Amphibious Entrepreneurs and the Emergence of Organizational Forms." *Strategic Entrepreneurship Journal* 6: 94–115.

Powell, Walter W./Kurt W. Sandholtz. 2012b. "Chance, Nécessité, et Naïveté," in John F. Padgett/Walter W. Powell (eds.), *The Emergence of Organizations and Markets*. Princeton, NJ: Princeton University Press: 379–433.

Powell, Walter W./Douglas White/Kenneth Koput/Jason Owen-Smith. 2005. "Network Dynamics and Field Evolution: The Growth of Interorganizational Collaboration in the Life Sciences." *American Journal of Sociology* 110: 1132–1205.

PR Newswire. 1990. "IDEC Pharmaceuticals to Establish Japanese Subsidiary." PR Newswire, February 26, 1990.

PR Newswire. 1992. "Biomira Inc. Announces Results of Phase I Clinical Trials of Its Theratope Products." PR Newswire, July 23, 1992.

PR Newswire. 1996. "Agouron Pharmaceuticals Reports Recent Results from Testing of Three Anti-Cancer Drugs." PR Newswire, April 22, 1996.

Pusztai, Lajos. 2008. "Individualized Therapy of Breast Cancer: Are We There Yet?" *Personalized Medicine* 5: 557–559.

Putnam, Linda L./Gail T. Fairhurst. 2001. "Discourse Analysis in Organizations," in Fredric M. Jablin/Linda L. Putnam (eds.), *The New Handbook of Organizational Communication: Advances in Theory, Research, and Methods*. Newbury Park, CA: SAGE: 78–136.

Putnam, Linda L./Suzanne Boys. 2006. "Revisiting Metaphors of Organizational Communication," in Steward R. Clegg/Cynthia Hardy/Thomas B. Lawrence/Walter Nord (eds.), *The SAGE Handbook of Organization Studies*. London: SAGE: 541–577.

Putnam, Linda L./François Cooren. 2004. "Alternative Perspectives on the Role of Text and Agency in Constituting Organizations." *Organization* 11: 323–333.

Quinn, Kevin M./Burt L. Monroe/Michael Colaresi/Michael H. Crespin/Dragomir R. Radev. 2010. "How to Analyze Political Attention with Minimal Assumptions and Costs." *American Journal of Political Science* 54: 209–228.

Rabeharisoa, Vololona/Michel Callon. 2002. "The Involvement of Patients' Associations in Research." *International Social Science Journal* 54: 57.

Rajan, Kaushik Sunder. 2006. *Biocapital: The Constitution of Postgenomic Life*. London and Durham, NC: Duke University Press.

Ramage, Daniel/Evan Rosen/Jason Chuang/Christopher D. Manning/Daniel A. McFarland. 2009. "Topic Modeling for the Social Sciences." *Workshop on Applications for Topic Models, NIPS* http://vis.stanford.edu/papers/topic-modeling-social-sciences.

Rao, Hayagreeva/Philippe Monin/Rodolphe Durand. 2003. "Institutional Change in Toque Ville: Nouvelle Cuisine as an Identity Movement in French Gastronomy." *American Journal of Sociology* 108: 795–843.

Ratner, Mark. 1990. "A New Era for Genentech, and So It Goes." *Nature Biotechnology* 8: 178–179.

Rauch, James E./Gary G. Hamilton. 2001. "Networks and Markets: Concepts for Bridging Disciplines," in James E. Rauch/Alessandra Casella (eds.), *Networks and Markets*. New York: SAGE: 1–29.
Rawlings, Craig M./Clayton Childress. 2021. "Measure Mohr Culture." *Poetics* 88: 101611.
Reichmann, Werner. 2013. "Epistemic Participation: How to Produce Knowledge about the Economic Future." *Social Studies of Science* 43: 852–877.
Riessman, Catherine K. 1993. *Narrative Analysis*. Thousand Oaks, CA: SAGE.
Rifkin, Jeremy. 1998. *The Biotech Century: Harnessing the Gene and Remaking the World*. London: Gollancz.
Robbins-Roth, Cynthia. 2000. *From Alchemy to IPO: The Business of Biotechnology*. Cambridge, MA: Perseus.
Rosa, José Antonio/Joseph F. Porac/Jelena Runser-Spanjol/Michael S. Saxon. 1999. "Sociocognitive Dynamics in a Product Market." *Journal of Marketing* 63: 64–77.
Ross, Jeffrey S./David P. Schenkein/Robert Pietrusko/Mark Rolfe/Gerald P. Linette/James Stec/Nancy E. Stagliano/Geoffrey S. Ginsburg/W. Fraser Symmans/Lajos Pusztai/Gabriel N. Hortobagyi. 2004. "Targeted Therapies for Cancer 2004." *American Journal of Clinical Pathology* 122: 598–609.
Roth, Camille/Nikita Basov. 2020. "The Socio-Semantic Space of John Mohr." *Poetics* 78: 101437–101437.
Roush, Wade. 1997. "On the Biotech Pharm, a Race to Harvest New Cancer Cures." *Science* 278: 1039–1040.
Roy, Roopali/Jiang Yang/Marsha A Moses. 2009. "Matrix Metalloproteinases as Novel Biomarkers and Potential Therapeutic Targets in Human Cancer." *Journal of Clinical Oncology* 27: 5287–5297.
Rozenblatt-Rosen, Orit/Aviv Regev/Philipp Oberdoerffer/Tal Nawy/Anna Hupalowska/Jennifer E. Rood/Orr Ashenberg/Ethan Cerami/Robert J. Coffey/Emek Demir/Li Ding/Edward D. Esplin/James M. Ford/Jeremy Goecks/Sharmistha Ghosh/Joe W. Gray/Justin Guinney/Sean E. Hanlon/Shannon K. Hughes/E. Shelley Hwang/Christine A. Iacobuzio-Donahue/Judit Jané-Valbuena/Bruce E. Johnson/Ken S. Lau/Tracy Lively/Sarah A. Mazzilli/Dana Pe'er/Sandro Santagata/Alex K. Shalek/Denis Schapiro/Michael P. Snyder/Peter K. Sorger/Avrum E. Spira/Sudhir Srivastava/Kai Tan/Robert B. West/Elizabeth H. Williams/Denise Aberle/Samuel I. Achilefu/Foluso O. Ademuyiwa/Andrew C. Adey/Rebecca L. Aft/Rachana Agarwal/Ruben A. Aguilar/Fatemeh Alikarami/Viola Allaj/Christopher Amos/Robert A. Anders/Michael R. Angelo/Kristen Anton/Orr Ashenberg/Jon C. Aster/Ozgun Babur/Amir Bahmani/Akshay Balsubramani/David Barrett/Jennifer Beane/Diane E. Bender/Kathrin Bernt/Lynne Berry/Courtney B. Betts/Julie Bletz/Katie Blise/Adrienne Boire/Genevieve Boland/Alexander Borowsky/Kristopher Bosse/Matthew Bott/Ed Boyden/James Brooks/Raphael Bueno/Erik A. Burlingame/Qiuyin Cai/Joshua Campbell/Wagma Caravan/Ethan Cerami/Hassan Chaib/Joseph M. Chan/Young Hwan Chang/Deyali Chatterjee/Ojasvi Chaudhary/Alyce A. Chen/Bob Chen/Changya Chen/Chia-hui Chen/Feng Chen/Yu-An Chen/Milan G. Chheda/Koei Chin/Roxanne Chiu/Shih-Kai Chu/Rodrigo Chuaqui/Jaeyoung Chun/Luis Cisneros/Robert

J. Coffey/Graham A. Colditz/Kristina Cole/Natalie Collins/Kevin Contrepois/Lisa M. Coussens/Allison L. Creason/Daniel Crichton/Christina Curtis/Tanja Davidsen/ Sherri R. Davies/Ino de Bruijn/Laura Dellostritto/Angelo De Marzo/Emek Demir/ David G. DeNardo/Dinh Diep/Li Ding/Sharon Diskin/Xengie Doan/Julia Drewes/ Stephen Dubinett/Michael Dyer/Jacklynn Egger/Jennifer Eng/Barbara Engelhardt/ Graham Erwin/Edward D. Esplin/Laura Esserman/Alex Felmeister/Heidi S. Feiler/ Ryan C. Fields/Stephen Fisher/Keith Flaherty/Jennifer Flournoy/James M. Ford/ Angelo Fortunato/Allison Frangieh/Jennifer L. Frye/Robert S. Fulton/Danielle Galipeau/Siting Gan/Jianjiong Gao/Long Gao/Peng Gao/Vianne R. Gao/Tim Geiger/ Ajit George/Gad Getz/Sharmistha Ghosh/Marios Giannakis/David L. Gibbs/William E. Gillanders/Jeremy Goecks/Simon P. Goedegebuure/Alanna Gould/Kate Gowers/Joe W. Gray/William Greenleaf/Jeremy Gresham/Jennifer L. Guerriero/ Tuhin K. Guha/Alexander R. Guimaraes/Justin Guinney/David Gutman/Nir Hacohen/Sean Hanlon/Casey R. Hansen/Olivier Harismendy/Kathleen A. Harris/Aaron Hata/Akimasa Hayashi/Cody Heiser/Karla Helvie/John M. Herndon/Gilliam Hirst/ Frank Hodi/Travis Hollmann/Aaron Horning/James J. Hsieh/Shannon Hughes/ Won Jae Huh/Stephen Hunger/Shelley E. Hwang/Christine A. Iacobuzio-Donahue/ Heba Ijaz/Benjamin Izar/Connor A. Jacobson/Samuel Janes/Judit Jané-Valbuena/ Reyka G. Jayasinghe/Lihua Jiang/Brett E. Johnson/Bruce Johnson/Tao Ju/Humam Kadara/Klaus Kaestner/Jacob Kagan/Lukas Kalinke/Robert Keith/Aziz Khan/ Warren Kibbe/Albert H. Kim/Erika Kim/Junhyong Kim/Annette Kolodzie/Mateusz Kopytra/Eran Kotler/Robert Krueger/Kostyantyn Krysan/Anshul Kundaje/Uri Ladabaum/Blue B. Lake/Huy Lam/Rozelle Laquindanum/Ken S. Lau/Ashley M. Laughney/Hayan Lee/Marc Lenburg/Carina Leonard/Ignaty Leshchiner/Rochelle Levy/ Jerry Li/Christine G. Lian/Kian-Huat Lim/Jia-Ren Lin/Yiyun Lin/Qi Liu/Ruiyang Liu/Tracy Lively/William J. R. Longabaugh/Teri Longacre/Cynthia X. Ma/Mary Catherine Macedonia/Tyler Madison/Christopher A. Maher/Anirban Maitra/Netta Makinen/Danika Makowski/Carlo Maley/Zoltan Maliga/Diego Mallo/John Maris/ Nick Markham/Jeffrey Marks/Daniel Martinez/Robert J. Mashl/Ignas Masilionais/ Jennifer Mason/Joan Massagué/Pierre Massion/Marissa Mattar/Richard Mazurchuk/Linas Mazutis/Sarah A. Mazzilli/Eliot T. McKinley/Joshua F. McMichael/Daniel Merrick/Matthew Meyerson/Julia R. Miessner/Gordon B. Mills/Meredith Mills/ Suman B. Mondal/Motomi Mori/Yuriko Mori/Elizabeth Moses/Yael Mosse/Jeremy L. Muhlich/George F. Murphy/Nicholas E. Navin/Tal Nawy/Michel Nederlof/Reid Ness/Stephanie Nevins/Milen Nikolov/Ajit Johnson Nirmal/Garry Nolan/Edward Novikov/Philipp Oberdoerffer/Brendan O'Connell/Michael Offin/Stephen T. Oh/ Anastasiya Olson/Alex Ooms/Miguel Ossandon/Kouros Owzar/Swapnil Parmar/ Tasleema Patel/Gary J. Patti/Dana Pe'er/Itsik Pe'er/Tao Peng/Daniel Persson/Marvin Petty/Hanspeter Pfister/Kornelia Polyak/Kamyar Pourfarhangi/Sidharth V. Puram/ Qi Qiu/Álvaro Quintanal-Villalonga/Arjun Raj/Marisol Ramirez-Solano/Rumana Rashid/Ashley N. Reeb/Aviv Regev/Mary Reid/Adam Resnick/Sheila M. Reynolds/ Jessica L. Riesterer/Scott Rodig/Joseph T. Roland/Sonia Rosenfield/Asaf Rotem/ Sudipta Roy/Orit Rozenblatt-Rosen/Charles M. Rudin/Marc D. Ryser/Sandro

Santagata/Maria Santi-Vicini/Kazuhito Sato/Denis Schapiro/Deborah Schrag/Nikolaus Schultz/Cynthia L. Sears/Rosalie C. Sears/Subrata Sen/Triparna Sen/Alex Shalek/Jeff Sheng/Quanhu Sheng/Kooresh I. Shoghi/Martha J. Shrubsole/Yu Shyr/Alexander B. Sibley/Kiara Siex/Alan J. Simmons/Dinah S. Singer/Shamilene Sivagnanam/Michal Slyper/Michael P. Snyder/Artem Sokolov/Sheng-Kwei Song/Peter K. Sorger/Austin Southard-Smith/Avrum Spira/Sudhir Srivastava/Janet Stein/Phillip Storm/Elizabeth Stover/Siri H. Strand/Timothy Su/Damir Sudar/Ryan Sullivan/Lea Surrey/Mario Suvà/Kai Tan/Nadezhda V. Terekhanova/Luke Ternes/Lisa Thammavong/Guillaume Thibault/George V. Thomas/Vésteinn Thorsson/Ellen Todres/Linh Tran/Madison Tyler/Yasin Uzun/Anil Vachani/Eliezer Van Allen/Simon Vandekar/Deborah J. Veis/Sébastien Vigneau/Arastoo Vossough/Angela Waanders/Nikhil Wagle/Liang-Bo Wang/Michael C. Wendl/Robert West/Elizabeth H. Williams/Chi-yun Wu/Hao Wu/Hung-Yi Wu/Matthew A. Wyczalkowski/Yubin Xie/Xiaolu Yang/Clarence Yapp/Wenbao Yu/Yinyin Yuan/Dadong Zhang/Kun Zhang/Mianlei Zhang/Nancy Zhang/Yantian Zhang/Yanyan Zhao/Daniel Cui Zhou/Zilu Zhou/Houxiang Zhu/Qin Zhu/Xiangzhu Zhu/Yuankun Zhu/Xiaowei Zhuang. 2020. "The Human Tumor Atlas Network: Charting Tumor Transitions across Space and Time at Single-Cell Resolution." *Cell* 181: 236–249.

Rule, Alix/Jean-Philippe Cointet/Peter S. Bearman. 2015. "Lexical Shifts, Substantive Changes, and Continuity in State of the Union Discourse, 1790–2014." *Proceedings of the National Academy of Sciences* 112: 10837–10844.

Russell, Sabin 1990. "Genentech Deal Makes Roche Biotech Giant." *San Francisco Chronicle*, February 6, 1990.

Salgado, Roberto/Helen Moore/John W. M. Martens/Tracy Lively/Shakun Malik/Ultan McDermott/Stefan Michiels/Jeffrey A. Moscow/Sabine Tejpar/Tawnya McKee/Denis Lacombe. 2018. "Steps Forward for Cancer Precision Medicine." *Nature Reviews Drug Discovery* 17: 1–2.

Santos, Filipe M./Kathleen M. Eisenhardt. 2009. "Constructing Markets and Shaping Boundaries: A Model of Entrepreneurial Action in Nascent Fields." *Academy of Management Journal* 52: 643–671.

Savage, Mike/Roger Burrows. 2007. "The Coming Crisis of Empirical Sociology." *Sociology: The Journal of the British Sociological Association* 41: 885–899.

Sawyer, R. Keith. 2001. "Emergence in Sociology: Contemporary Philosophy of Mind and Some Implications for Sociological Theory." *American Journal of Sociology* 107: 551–585.

Schilsky, Richard L. 2010. "Personalized Medicine in Oncology: The Future Is Now." *Nature Reviews Drug Discovery* 9: 363–366.

Schmid, Helmut. 1995. "Treetagger: A Language Independent Part-of-Speech Tagger." Institut für Maschinelle Sprachverarbeitung, Universität Stuttgart, http://www.cis.uni-muenchen.de/~schmid/tools/TreeTagger/.

Schnegg, Michael/H. Russell Bernard. 1996. "Words as Actors: A Method for Doing Semantic Network Analysis." *Field Methods* 8: 7–10.

Schneiberg, Marc/Elisabeth S. Clemens. 2006. "The Typical Tools for the Job: Research Strategies in Institutional Analysis." *Sociological Theory* 24: 195–227.

Schnitt, Stuart J./Sunil R. Lakhani. 2014. "Breast Cancer," in Bernard W. Stewart/Christopher P. Wild (eds.), *World Cancer Report 2014*. Lyon: IARC: 362–373.

Schoeneborn, Dennis/Steffen Blaschke/François Cooren/Robert D. McPhee/David Seidl/James R. Taylor. 2014. "The Three Schools of CCO Thinking: Interactive Dialogue and Systematic Comparison." *Management Communication Quarterly* 28: 285–316.

Schreyögg, Georg/Jochen Koch. 2005. *Knowledge Management and Narratives*. Berlin: Erich Schmidt Verlag.

Schubert, Cornelius/Jörg Sydow/Arnold Windeler. 2013. "The Means of Managing Momentum: Bridging Technological Paths and Organisational Fields." *Research Policy* 42: 1389–1405.

Searle, John R. 1975. "The Logical Status of Fictional Discourse." *New Literary History* 6: 328.

Seget, Steven. 2004. *The Cancer Market Outlook to 2009*. London: Business Insights.

Selin, Cynthia. 2007. "Expectations and the Emergence of Nanotechnology." *Science, Technology, & Human Values* 32: 196–220.

Selin, Cynthia. 2008. "The Sociology of the Future: Tracing Stories of Technology and Time." *Sociology Compass* 2: 1878–1895.

Service, Robert F. 2004. "Surviving the Blockbuster Syndrome." *Science* 303: 1796–1799.

Shane, Scott/Sankaran Venkataraman. 2000. "The Promise of Entrepreneurship as a Field of Research." *Academy of Management Review* 25: 217–226.

Sharpless, Norman E./Dinah S. Singer. 2021. "Progress and Potential: The Cancer Moonshot." *Cancer Cell* 39: 889–894.

Shiller, Robert J. 2019. *Narrative Economics: How Stories Go Viral & Drive Major Economic Events*. Princeton, NJ: Princeton University Press.

Silber, B. Michael. 2001. "Pharmacogenomics, Biomarkers, and the Promise of Personalized Medicine." *Drugs and the Pharmaceutical Sciences* 113: 11–32.

Sillince, John/Paula Jarzabkowski/Duncan Shaw. 2012. "Shaping Strategic Action Through the Rhetorical Construction and Exploitation of Ambiguity." *Organization Science* 23: 630–650.

Simmel, Georg. 1992 [1908]. "Die Kreuzung sozialer Kreise," in Otthein Rammstedt (ed.), *Soziologie: Untersuchungen über die Formen der Vergesellschaftung*. Frankfurt: Suhrkamp Band 11: 456–511.

Sinn, Hans-Peter/Hans Kreipe. 2013. "A Brief Overview of the WHO Classification of Breast Tumors, 4th Edition, Focusing on Issues and Updates from the 3rd Edition." *Breast Care* 8: 149–154.

Skeel, Shirley. 1994. "Bio Pins Hopes on Cancer Drug." *Daily Mail*, March 10, 1994.

Slamon, Dennis J./William Godolphin/Lovell A. Jones/John A. Holt/Steven G. Wong/Duane E. Keith/Wendy J. Levin/Susan G. Stuart/Judy Udove/Axel Ullrich. 1989. "Studies of the HER-2/neu Proto-oncogene in Human Breast and Ovarian Cancer." *Science* 244: 707–712.

Slamon, Dennis J./Gary M. Clark/Steven G. Wong/Wendy J. Levin/Axel Ullrich/William L. McGuire. 1987. "Human Breast Cancer: Correlation of Relapse and Survival with Amplification of the HER-2/neu Oncogene." *Science* 235: 177–182.

Slavich, Barbara/Silviya Svejenova/M. Pilar Opazo/Gerardo Patriotta. 2020. "Politics of Meaning in Categorizing Innovation: How Chefs Advanced Molecular Gastronomy by Resisting the Label." *Organization Studies* 41: 267–290.

Smith, Kendall A. 1989. "Interleukin Futures." *Nature Biotechnology* 7: 661–667.

Smith, Tammy. 2007. "Narrative Boundaries and the Dynamics of Ethnic Conflict and Conciliation." *Poetics* 35: 22–46.

Sosa, M. Lourdes. 2011. "From Old Competence Destruction to New Competence Access: Evidence from the Comparison of Two Discontinuities in Anticancer Drug Discovery." *Organization Science* 22: 1500–1516.

Spence, Michael. 1973. "Job Market Signaling." *Quarterly Journal of Economics* 87: 355–374.

Spence, Michael. 1974. *Market Signaling: Informational Transfer in Hiring and Related Screening Processes*. Cambridge, MA: Harvard University Press.

Spence, Michael. 2002. "Signaling in Retrospect and the Informational Structure of Markets." *American Economic Review* 92: 434–459.

Stark, David. 2009. *The Sense of Dissonance: Accounts of Worth in Economic Life*. Princeton, NJ: Princeton University Press.

Stehelin, Dominique/Harold E. Varmus/J. Michael Bishop/Peter K. Vogt. 1976. "DNA Related to the Transforming Gene(s) of Avian Sarcoma Viruses Is Present in Normal Avian DNA." *Nature* 260: 170–173.

Stevenson, Tom/Tom Wilkie. 1995. "City Frenzy over 'Cure for Cancer.'" *The Independent*, 2. December 1995, 1.

Steyvers, Mark/Tom Griffiths. 2007. "Probabilistic Topic Models," in Thomas K. Landauer/Danielle S. McNamara/Simon Dennis/Walter Kintsch (eds.), *Handbook of Latent Semantic Analysis*. Oxon: Psychology Press: 424–448.

Stinchcombe, Arthur. 1965. "Social Structures and Organizations," in James March (ed.), *Handbook of Organizations*. Chicago: University of Chicago Press: 142–193.

Stovel, Katherine/Daniel Koski-Karell. 2015. "Narrative Networks," in James D. Wright (ed.), *International Encyclopedia of the Social & Behavioral Sciences*. Oxford: Elsevier. 5: 211–217.

Stovel, Katherine/Michael Savage/Peter S. Bearman. 1996. "Ascription into Achievement: Models of Career Systems at Lloyds Bank, 1980–1970." *American Journal of Sociology* 102: 358–399.

Strebhardt, Klaus/Axel Ullrich. 2008. "Paul Ehrlich's Magic Bullet Concept: 100 Years of Progress." *Nature Reviews Cancer* 8: 473–480.

Styhre, Alexander. 2011. "Competing Institutional Logics in the Biopharmaceutical Industry: The Move Away from the Small Molecules Therapies Model in the Post-Genomic Era." *Creativity and Innovation Management* 20: 311–329.

Suddaby, Roy/William M. Foster/Chris Quinn Trank. 2010. "Rhetorical History as a Source of Competitive Advantage." *Globalization of Strategy Research* 27: 147–173.

Sun, Marjorie. 1981. "Interferon: No Magic Bullet against Cancer." *Science* 212: 141–142.
Sung, Hyuna/Jacques Ferlay/Rebecca L. Siegel/Mathieu Laversanne/Isabelle Soerjomataram/Ahmedin Jemal/Freddie Bray. 2021. "Global Cancer Statistics 2020: GLOBOCAN Estimates of Incidence and Mortality Worldwide for 36 Cancers in 185 Countries." *CA: A Cancer Journal for Clinicians*, https://doi.org/10.3322/caac.21660.
Swan, Jacky/Anna Goussevskaia/Sue Newell/Maxine Robertson/Mike Bresnen/Ademola Obembe. 2007. "Modes of Organizing Biomedical Innovation in the UK and US and the Role of Integrative and Relational Capabilities." *Research Policy* 36: 529–547.
Swidler, Ann. 1986. "Culture in Action: Symbols and Strategies." *American Sociological Review* 51: 273–286.
Syed, Basharut A. 2015. "The Breast Cancer Market." *Nature Reviews Drug Discovery* 14: 233–234.
Tafuri, Giovanni/P. Stolk/Francesco Trotta/M. Putzeist/Hubert G. M. Leufkens/Richard Laing/M. De Allegri. 2014. "How Do the EMA and FDA Decide Which Anticancer Drugs Make It to the Market? A Comparative Qualitative Study on Decision Makers' Views." *Annals of Oncology* 25: 265–269.
Tan, Puay Hoon/Ian Ellis/Kimberly Allison/Edi Brogi/Stephen B. Fox/Sunil Lakhani/Alexander J. Lazar/Elizabeth A. Morris/Aysegul Sahin/Roberto Salgado/Anna Sapino/Hironobu Sasano/Stuart Schnitt/Christos Sotiriou/Paul van Diest/Valerie A. White/Dilani Lokuhetty/Ian A. Cree/for the WHO Classification of Tumours Editorial Board. 2020. "The 2019 World Health Organization Classification of Tumours of the Breast." *Histopathology* 77: 181–185.
Tarde, Gabriel. 1962. *The Laws of Imitation*. 2nd ed. Gloucester, MA: Peter Smith.
Tavory, Iddo/Nina Eliasoph. 2013. "Coordinating Futures: Toward a Theory of Anticipation." *American Journal of Sociology* 118: 908–942.
Taylor, James R./Elizabeth J. van Every. 2000. *The Emergent Organization: Communication as Its Site and Surface*. Mahwah, NJ: Lawrence Erlbaum.
Taylor, James R./Elizabeth J. van Every. 2011. *The Situated Organization. Case Studies in the Pragmatics of Communication Research*. New York: Routledge.
Taylor, Paul. 1998. "In Search of the Cure." *The Globe and Mail*, October 13, 1998.
Teitelman, Robert. 1989. *Gene Dreams: Wall Street, Academia, and the Rise of Biotechnology*. New York: Basic Books.
Tellmann, Ute. 2003. "The Truth of the Market." *Distinktion: Scandinavian Journal of Social Theory* 4: 49–63.
The New York Times. 1998. "Genentech Earnings Double for Quarter." *The New York Times*, October 15, 1998.
The Times. 1996. "Merlin Ventures." *The Times*, November 5, 1996.
Thévenot, Laurent. 2001. "Organized Complexity: Conventions of Coordination and the Composition of Economic Arrangements." *European Journal of Social Theory* 4: 405–425.
Thévenot, Laurent/Michael Moody/Claudette Lafaye. 2000. "Forms of Valuing Nature: Arguments and Modes of Justification in French and American Environmental

Disputes," in Michèle Lamont/Laurent Thévenot (eds.), *Rethinking Comparative Cultural Sociology: Repertoires of Evaluation in France and the United States*. Cambridge: Cambridge University Press: 229–272.

Thornton, Patricia H./William Ocasio/Michael Lounsbury. 2012. *The Institutional Logics Perspective: A New Approach to Culture, Structure, and Process*. Oxford: Oxford University Press.

Thornton, Patricia H. 2004. *Markets from Culture: Institutional Logics and Organizational Decisions in Higher Education Publishing*. Stanford, CA: Stanford Business Books.

Tilly, Charles. 2006. *Why?* Princeton, NJ: Princeton University Press.

Timmermann, Carsten. 2014. *A History of Lung Cancer: The Recalcitrant Disease*. Basingstoke, UK: Palgrave Macmillan.

Toronto Star. 2003. "Biomira Shares Fall on Study Results." *Toronto Star*, June 18, 2003.

Trotta, Francesco/Hubert G. M. Leufkens/Jan H. M. Schellens/Richard Laing/Giovanni Tafuri. 2011. "Evaluation of Oncology Drugs at the European Medicines Agency and US Food and Drug Administration: When Differences Have an Impact on Clinical Practice." *Journal of Clinical Oncology* 29: 2266–2272.

Tutton, Richard. 2011. "Promising Pessimism: Reading the Futures to Be Avoided in Biotech." *Social Studies of Science* 41: 411–429.

Ullrich, Axel/Lisa Coussens/Joel S. Hayflick/Thomas J. Dull/Alane Gray/A. W. Tam/J. Lee/Y. Yarden/Towia A. Libermann/Joseph Schlessinger/Julian Downward/E. L. V. Mayes/N. Whittle/M. D. Waterfield/P. H. Seeburg. 1984. "Human Epidermal Growth Factor Receptor cDNA Sequence and Aberrant Expression of the Amplified Gene in A431 Epidermoid Carcinoma Cells." *Nature* 309: 418–425.

Uzzi, Brian. 1996. "The Sources and Consequences of Embeddedness for the Economic Performance of Organizations: The Network Effect." *American Sociological Review* 61: 674–698.

Uzzi, Brian. 1997. "Social Structure and Competition in Interfirm Networks: The Paradox of Embeddedness." *Administrative Science Quarterly* 42: 35–67.

Vaara, Eero/Janne Tienari. 2011. "On the Narrative Construction of Multinational Corporations: An Antenarrative Analysis of Legitimation and Resistance in a Cross-Border Merger." *Organization Science* 22: 370–390.

Van de Ven, Andrew H./Douglas Polley/Raghu Garud/Sankaran Venkataraman. 1999. *The Innovation Journey*. Oxford: Oxford University Press.

van Dijk, Teun A. 2001. "Critical Discourse Analysis," in Deborah Schiffren/Deborah Tannen/Heidi E. Hamilton (eds.), *The Handbook of Discourse Analysis*. Oxford: Blackwell: 352–371.

van Lente, Harro/Arie Rip. 1998. "Expectations in Technological Developments: An Example of Prospective Structures to be Filled In by Agency," in Cornelis Disco/Barend van der Meulen (eds.), *Getting New Technologies Together: Studies in Making Sociotechnical Order*. Berlin, Boston: De Gruyter: 203–230.

van Lente, Harro. 1993. *Promising Technology: The Dynamics of Expectations in Technological Developments*. Enschede, Netherlands: Universiteit Twente.

Van Poznak, Catherine/Andrew D. Seidman/R. Bertino Joseph. 2002. "Breast Cancer." *Encyclopedia of Cancer*. New York: Academic Press: 287–299.

Vasella, Daniel/Robert Slater. 2003. *Magic Cancer Bullet: How a Tiny Orange Pill Is Rewriting Medical History*. New York: HarperBusiness.

Venturini, Tommaso/Nicolas Baya Laffite/Jean-Philippe Cointet/Ian Gray/Vinciane Zabban/Kari De Pryck. 2014. "Three Maps and Three Misunderstandings: A Digital Mapping of Climate Diplomacy." *Big Data & Society* 2, http://bds.sagepub.com/content/1/2/2053951714543804.

Vergne, Jean-Philippe. 2012. "Stigmatized Categories and Public Disapproval of Organizations: A Mixed-Methods Study of the Global Arms Industry, 1996–2007." *Academy of Management Journal* 55: 1027–1052.

Vergne, Jean-Philippe/Gautam Swain. 2017. "Categorical Anarchy in the UK? The British Media's Classification of Bitcoin and the Limits of Categorization," in Rodolphe Durand/Nina Granqvist/Anna Tyllström (eds.), *From Categories to Categorization: Studies in Sociology, Organizations, and Strategy at the Crossroads*. Bingley, UK: Emerald Publishing 51: 187–222.

Wagener, Damianus Johannes Theodorus. 2009. *The History of Oncology*. Houten, Netherlands: Springer.

Wallach, Hanna M./David M. Mimno/Andrew McCallum. 2009. "Rethinking LDA: Why Priors Matter." *Advances in Neural Information Processing Systems* 22, http://papers.nips.cc/paper/3854-rethinking-lda-why-priors-matter.pdf.

Watts, Duncan J. 2013. "Computational Social Science." *The Bridge* 43: 5–11.

Weber, Klaus/Kathryn L. Heinze/Michaela DeSoucey. 2008. "Forage for Thought: Mobilizing Codes in the Movement for Grass-Fed Meat and Dairy Products." *Administrative Science Quarterly* 53: 529–567.

Weick, Karl E.. 1995. *Sensemaking in Organizations*. Thousand Oaks, CA: SAGE.

Weick, Karl E. 1993. "The Collapse of Sensemaking in Organizations: The Mann Gulch Disaster." *Administrative Science Quarterly* 38: 628–652.

Weick, Karl E./Kathleen M. Sutcliffe/David Obstfeld. 2005. "Organizing and the Process of Sensemaking." *Organization Science* 16: 409–421.

Weinberg, Robert A. 1996. "How Cancer Arises." *Scientific American* 275: 62–71.

Weiner, Louis M./Joseph I. Clark/Monica Davey/Wei S. Li/Irma Garcia de Palazzo/David B. Ring/R. Katherine Alpaugh. 1995. "Phase I Trial of 2B1, a Bispecific Monoclonal Antibody Targeting c-erbB-2 and FcγRIII." *Cancer Research* 55: 4586–4593.

Wetzel, Dietmar J. 2013. "Dispositive, Discourse and the Economy—Conceptual Reflections with Regard to a Sociology of Competition." *Economic Sociology: The European Electronic Newsletter* 14: 10–16.

Wherry, Frederick F. 2012. *The Culture of Markets*. Cambridge: Polity.

Wherry, Frederick F. 2014. "Analyzing the Culture of Markets." *Theory and Society* 43: 421–436.

White, Douglas R./Jason Owen-Smith/James Moody/Walter W. Powell. 2004. "Networks, Fields and Organizations: Micro-Dynamics, Scale, and Cohesive Embeddings." *Computational & Mathematical Organization Theory* 10: 95–117.

White, Harrison C. 1981. "Where Do Markets Come From?" *American Journal of Sociology* 87: 517–547.
White, Harrison C. 1992. *Identity and Control: A Structural Theory of Social Action*. Princeton, NJ: Princeton University Press.
White, Harrison C. 1993. "Markets in Production Networks," in Richard Swedberg (ed.), *Explorations in Economic Sociology*. New York: Russell Sage Foundation: 161–175.
White, Harrison C. 1995. "Network Switchings and Bayesian Forks: Reconstructing the Social and Behavioral Sciences." *Social Research* 62: 1035–1063.
White, Harrison C. 2000. "Modeling Discourse in and around Markets." *Poetics* 27: 117–133.
White, Harrison C. 2002. *Markets from Networks: Socioeconomic Models of Production*. Princeton, NJ: Princeton University Press.
White, Harrison C. 2003. "Meanings Out of Ambiguity from Switchings, with Grammar as One Trace: Directions for a Sociology of Language." The Cultural Turn at UC Santa Barbara: Instituting and Institutions. University of California, Santa Barbara.
White, Harrison C. 2008. *Identity and Control: How Social Formations Emerge*. Princeton, NJ: Princeton University Press.
White, Harrison C./Scott Boorman/Ronald L. Breiger. 1976. "Social Structure from Multiple Networks. I. Blockmodels of Roles and Positions." *American Journal of Sociology* 81: 730–779.
White, Harrison C./Robert Eccles. 1987. "Producers' Markets," in John Eatwell (ed.), *The New Palgrave Dictionary of Economic Theory and Doctrine*. London: Macmillan: 984–986.
White, Harrison C./Jan Fuhse/Matthias Thiemann/Larissa Buchholz. 2007. "Networks and Meaning: Styles and Switchings." *Soziale Systeme* 13: 514–526.
White, Harrison C./Frédéric Godart. 2007. "Stories from Identity and Control." *Sociologica* 3, http://www.sociologica.mulino.it/doi/10.2383/25960.
White, Hayden. 1980. "The Value of Narrativity in the Representation of Reality." *Critical Inquiry* 7: 5–27.
Whittington, Kjersten Bunker/Jason Owen-Smith/Walter W. Powell. 2009. "Networks, Propinquity, and Innovation in Knowledge-Intensive Industries." *Administrative Science Quarterly* 54: 90–122.
Wilcock, Paul/Rachel M. Webster. 2021. "The Breast Cancer Drug Market." *Nature Reviews Drug Discovery* 20: 339–340.
Wimmer, Andreas. 2008. "The Making and Unmaking of Ethnic Boundaries: A Multilevel Process Theory." *American Journal of Sociology* 113: 970–1022.
Winer, Arthur/Sylvia Adams/Paolo Mignatti. 2018. "Matrix Metalloproteinase Inhibitors in Cancer Therapy: Turning Past Failures into Future Successes." *Molecular Cancer Therapeutics* 17: 1147–1155.
Wodak, Ruth (ed.), 2013. *Critical Discourse Analysis*. Los Angeles: SAGE.
Yao, Lixia/James A. Evans/Andrey Rzhetsky. 2010. "Novel Opportunities for Computational Biology and Sociology in Drug Discovery: Corrected Paper." *Trends in Biotechnology* 28: 161–170.

Zehr, Leonard. 2002. "Biomira Shares Plummet 42%." *The Globe and Mail*, September 20, 2002.
Zelizer, Viviana A. 1996. "Payments and Social Ties." *Sociological Forum* 11: 481–495.
Zelizer, Viviana A. 1978. "Human Values and the Market: The Case of Life Insurance and Death in 19th-Century America." *American Journal of Sociology* 84: 591–610.
Zelizer, Viviana A. 1989. "The Social Meaning of Money: 'Special Monies.'" *American Journal of Sociology* 95: 342–377.
Zerubavel, Eviatar. 1996. "Lumping and Splitting: Notes on Social Classification." *Sociological Forum* 11: 42–33.
Zerubavel, Eviatar. 1997. *Social Mindscapes: An Invitation to Cognitive Sociology*. Cambridge, MA: Harvard University Press.
Zhao, Eric Yanfei/Masakazu Ishihara/Michael Lounsbury. 2013. "Overcoming the Illegitimacy Discount: Cultural Entrepreneurship in the US Feature Film Industry." *Organization Studies* 34: 1747–1776.
Zhao, Wei. 2005. "Understanding Classifications: Empirical Evidence from the American and French Wine Industries." *Poetics* 33: 179–200.
Zuckerman, Ezra W. 1999. "The Categorical Imperative: Securities Analysts and the Illegitimacy Discount." *American Journal of Sociology* 104: 1398–1438.

Index

Activase, 56
Actor-network theory (ANT), 11–15, 24, 34, 150nn8–9, 151n5
Actors: competition and, 21–22 (*see also* competition); emergence and, 18–45; expectations and, 58–60, 69–71; innovation and, 57; making sense of market and, 74, 84, 87; meaning-making and, 89, 97, 123; methodology and, 139, 159n3; pipelines and, 20, 65–66, 68, 75, 81, 86, 99, *101*, *103*, 104, 106, 123; sociocognitive, 2, 6, 12–13, 16, 19, 25, 27, 37, 41, 124, 128, 130, 132; sociomaterial, 2, 6, 12–13, 16, 19, 25, 27, 37, 44, 124, 130, 132; stories and, 1–18 (*see also* stories); text as data and, 91, 134, 139–40
Adjuncts, 69, 75, *76*, *78*, 82
Agouron Pharmaceuticals, 66, 68
Algorithms: categories and, 90–96, 109, 111–12, 118–19, 147, 156n11; formal analyses and, 145–48; Latent Dirichlet Allocation (LDA), 91–97, 146, 156n3; Louvain community detection, 111–12, 118–19, 147; meaning-making and, 111–12, 118–19, 147; networks and, 90, 110, 118, 142, 147, 150n3, 150n9; topic models and, 90–96, 109
Ambiguity: expectations and, 71; uncertainty and, 5, 19, 28, 34, 37, 40–41, 45, 71, 128, 152n18
Amgen, 53, 56

Angiogenesis: blood and, 59, 70, 77, 79–83, *105*, 116, 155n13; expectations and, 59, 70; making sense of market and, 77–83; matrix metalloproteinase (MMP) inhibitors and, 59, 70, 77; meaning-making and, *105*, 116
Antibodies: expectations and, 59–63, 67; making sense of market and, 76–86; meaning-making and, 100, *101*, *105*, 106–8, 116, 123; monoclonal (mAbs), 46, 55, 59, 63, 67, 76–86, 100, *101*, *105*, 106, *108*, 116, 123, 130, 147, 153n11, 155n11, 157n6; therapies and, 46, 55, 156n6
Antigens: expectations and, 59–62; making sense of market and, 79, 86; meaning-making and, 97, *101*, 119, 123; therapies and, 55
Antisoma, 67, 70
AstraZeneca, 76–77, 81, 157n9, 158n17
Avastin, 79–83, 97, 107–8, *108*

Batimastat: British Biotech and, 63–67, 100, *105*, 106–10; expectations and, 63–67; market share of, 64; meaning-making and, 100, *105*, 106–10
Bazell, Robert, 58, 154n9
Beckert, Jens, 5–6, 38–40, 125
Bevacizumab: as Avastin, 79–83, 97, *107*, *108*; meaning-making and, 114, 116; stories and, 133, 136

201

Beyond the Blockbuster Drug (Mertens), 81–82
Bioentrepreneur magazine, 140, 158n2
Biogen, 53, 55, 153n10, 154n10
Biomarkers, 82, 87, 130, 156n5, 157n3
Biomira, 60–63, 66–68, 70, 155n14
Biotechnology: clinical trials and, 2 (*see also* clinical trials); competition and, 2, 53, 58, 60, 65, 69–71, 142, 153n19; entrepreneurship and, 10, 53, 140, 158n2; expectations and, 58–72, 153n1, 154n3, 154nn7–8, 154n10, 155n12; high stakes of, 11–12; innovation and, 1–2, 4, 7–11, 15–17, 46–47, 53–60, 67, 126, 128, 133, 140, 153n9; investors and, 1–2, 46, 53–57, 63–64, 71–72, 74, 98, 141, 158n1; knowledge creation and, 8–9; late 1980s treatments and, 53–57; making sense of market and, 73–74, 85–87; matrix metalloproteinase (MMP) inhibitors and, 63, 65, 70, 77; meaning-making and, 90, 96, 98, *101*, 102, 106, *108*, 123; methodology and, 139–43; mTOR protein and, 158n15; positionings and, 61–69; stories and, 2–4, 15, 58, 60, 69–71, 74, 87, 96, 123, 126–29, 133, 139, 153n9; as textual data source, 15; therapies and, 1–3, 5, 16, 46–47, 53–63, 67–74, 85–87, 90, *101*, *105*, 123, 127–29, 133, 139–40, 158n2
Biotechnology magazine, 65, 140, 158n2
Blackboxing, 150n8
Blockbuster drugs: biosimilars and, 157n4; expectations and, 63–65; making sense of market and, 75, 77, 80–81, 86–87; potential, 63; therapies and, 6, 63–65, 75, 77, 80–81, 86–87, 129–32, 157n4
Blood: Activase and, 56; angiogenesis and, 59, 70, 77, 79–83, *105*, 116, 155n13; cell mobilization and, 82; erythropoietin (EPO) and, 56, 76; expectations and, 59, 155n13; making sense of market and, 82–83; meaning-making and, *101*, *108*; therapies and, 47–49, 52, 56; vessel growth and, 52–53, 83, 158n15
Boltanski, Luc, 25, 27–28, 151n8
BRCA genes, 50, 82, 136–37, 158n17
Breast cancer: advanced, 47; as leading cause of death for women, 1, 47; categories and, 89–123 (*see also* categories); causes of, 47; characteristics of, 47; diagnosis of, 1, 47, 49, 82, 86–87, 102, 152n1, 154n9, 156n4, 157n3; environmental factors for, 47; expectations and, 58–72; history of treatment of, 48–53; increase of, 47; late 1980s treatments and, 53–57; lifestyle risk factors for, 47; meaning-making and, 89–123; metastasis and, 47, *101*, 137; oncogene and, 1, 52–53, 55, 59, 61, 153n8; stories and, 18 (*see also* stories); therapies for, 46–57 (*see also* therapies); transmission of, 47; triple-negative (TNBC), 136, 157n5
Breiger, Ronald, 11, 27, 35, 110, 156n10
Bristol-Myers Squibb (BMS), 60, 68, 75–77, 81
British Biotech: Batimastat and, 63–67, 100, *105*, 106–10; expectations and, 154n10; innovation and, 53, 63–68, 70; Marimastat and, 64–67, 100, 102–9; meaning-making and, 102, *105*, *108*, 123; stock prices and, 65, 154n10
Business Insights, 74–75, 77, 82–83, 86, 140, *141*

Callon, Michel, 24, 28, 44-45
Cancer Market Outlook to 2007 (Business Insights), 77
Cancer Market Outlook to 2009 (Business Insights), 74, 79
Cancer Market Outlook to 2016 (Business Insights), 82–83
Cancer Outlook 2000 (Business Insights), 74–75, 77
Cancer Research Data Commons (CRDC), 138
Cancer Research journal, 94

Index

Carcinogenesis, 52
Categories: algorithms and, 90–96, 109, 111–12, 118–19, 147, 156n1; classification and, 6–7, 41, 43, 74, 88, 134; clinical trials and, 6, 129; clusters and, 90–94, 97–100, 110–13, 116–22; collaboration and, 3, 5, 17, 73, 87, 89–90, 123, 128–29, 132–33; construction of, 3, 5, 7, 29–42, 89–90, 129; delineating process of, 7; emergence and, 4–5, 14–15, 19, 37–38, 41–45, 74–75, 84, 87, 89, 110–23, 128–29, 135; entrepreneurship and, 3; formal analyses and, 145–48; industry analysts and, 89, 96, 98–102, 109, 123; journalists and, 89–90, 93–94, 106–7, 108, 110, 123; Latent Dirichlet Allocation (LDA) and, 91–97, 146, 156n3; Louvain community detection algorithm and, 111–12, 118–19, 147; making sense of market and, 73–80, 84–88; market structure and, 4–6, 123, 132–33; meaning-making and, 89–123, 156n1; methodology and, 14, 90, 110, 133, 135, 140–42, 145–48; movement in, 110–23; over time, 89–123; scientists and, 93–97, *108*; semantic network analyses and, 89–90, 110–23; stories and, 3–5, 17, 19, 28–45, 74, 84, 87–90, 123, 126–36; therapies and, 4, 6, 15, 17, 19, 73–76, 80, 84–90, 95–97, 100, *101*, 109–10, 116–23, 126–33, 136, 140, 147, 157n12; topic model analyses and, 91–110
CDK4/6 inhibitors, 137
Cells: blocking signals of, 61–62; blood mobilization and, 82; division of, 49, 137, 158n12, 158n15; expectations and, 59–63, 67, 70, 155n11; extracellular environment and, 59, 63; HER2 protein and, 55 (*see also* HER2 protein); inhibitors and, 51, 59, 63, 80, 82, 97, *101*, *105*, 119, 137, 157n5, 158n12; intracellular environment and, 59–60, 158n15; making sense of market and, 79–82, 85–86, 155n3; malignant, 47, 80, 86; meaning-making and, 90, 96–97, *101*, *105*, 110, 113, 117, 119–22, 157n12; mTOR protein and, 97, 109, 116, 119–22, 137, 158n15; oncogenes and, 1, 52–53, 55, 59, 61, 153n8; signaling of, 52–53, 61, 96, *101*, *105*, 157n10, 158n15; stem, 113, 157n12; target, 46, 52, 55, 59, 61, 79–82, 85–86, 90, 96, *101*, *105*, 113, 117, 119–22, 137, 148, 157n5, 157n12; therapies and, 46–56; tumorous, 47 (*see also* tumors); tyrosine kinase receptors and, 61–62, 82, 97, 155n3
Cetus, 53, 55
Chemotherapy: expectations and, 62–63, 66–69, 153n7, 154n9; HER2 protein and, 63, 77, 83, 116, 119, 136, 154n9; innovation and, 46, 49–51, 55, 136–37, 157nn8–9, 157n12; invasive nature of, 1, 46; making sense of market and, 75–77, 80–85, 156n6; meaning-making and, 95, *96*, 102, 104–10, 113–16, 119, 122
Chiron, 53, 62, 67–68
Chiroscience, 65–70, *108*
Chronic myeloid leukemia (CML), 155n3, 156n4
Clinical Cancer journal, 94
Clinical trials: categories and, 6, 129; expectations and, 61–71, 154n9, 155n14; making sense of market and, 77, 82, 86, 155n2; meaning-making and, 94, 97, *105*, 107, *108*; product approval and, 2, 5, 62, 68–69, 108, *108*, 116, 133, 136, 154n9; stories and, 129, 137, 157n5; therapies and, 46, 50, 55–57, 152n1
Clusters: of agreement, 41; categories and, 90–94, 97–100, 110–13, 116–22; Louvain community detection algorithm and, 111–12, 118–19, 147; meaning-making and, 90–94, 97–100, 110–13, 116–22; semantic network analyses and, 110–13, 116–22; stories and, 41; topic models and, 91–94, 97–100

Cognition: complexity and, 7, 10–11; emergence and, 2, 5–6, 10–13, 16–19, 25–29, 37–38, 42, 45, 84, 124–25, 128, 132–34; expectations and, 59, 70–71; innovation and, 2, 4, 7, 10–13, 19–22, 25–30, 33–34, 37–45, 124, 128, 133; making sense of market and, 84; networks and, 2, 4, 10–13, 16, 19–22, 25, 27, 37, 44, 124, 128, 130–33; reflexive, 16, 27–29, 125; sociology and, 6, 11–13, 19–20, 27; stories and, 4–6, 16, 19, 26, 30, 33, 37–41, 44–45, 71, 84, 124, 128–33

Cultural analysis, 6–7, 11–15, 18–20, 127, 131, 134, 151n5

Collaboration: categories and, 3, 5, 17, 73, 87, 89–90, 123, 128–29, 132–33; expectations and, 17, 59–60, 71; identity and, 13; innovation and, 53–55; making sense of market and, 73, 81, 87, 156n5; meaning-making and, 89–90, 105, 108, 123; networks and, 3, 13, 17, 60, 89–90, 123, 128–29, 133, 138, 150n9; stories and, 23–24, 128–29, 132–33, 138, 151n8; therapies and, 53–55

Communication: as data, 35; expectations and, 59–60, 71; language and, 13 (see also language); making sense of market and, 84; meaning-making and, 98; projectivity and, 19, 37–38; stories and, 3, 13, 16, 30–38, 42–43, 125, 127, 149n4

Competition: biotechnology and, 2, 53, 58, 60, 65, 69–71, 142, 153n19; expectations and, 58, 60, 65, 69–71, 153n10; innovation and, 53–54; making sense of market and, 80; narrative, 4; pharmaceutical industry and, 53–54, 65; stories and, 4–5, 16, 21–24, 32, 40–45, 58, 60, 69–71, 125–28, 132, 135, 150n1, 151n8; therapies and, 53–54; White model and, 21–22

Computational social science, 35, 90–93, 134, 142

Control: establishing, 12; expectations and, 63, 68; *Identity and Control*, 23, 131; stories and, 23, 29–30, 34, 131, 158n15

CorText, 146–48, 156

Decision-making, 2, 5, 16, 21, 126, 139

Diagnostics: FDA and, 86–87, 154n9; innovation and, 47, 49; specifics of, 102; targeted, 82; therapies and, 1, 47, 49, 82, 86–87, 102, 152n1, 154n9, 156n4, 157n3

DiMaggio, Paul, 8, 11, 15, 20, 91–93, 151n13

DNA, 52–55, 147n5

Drugs: blockbuster, 63 (see also blockbuster drugs); clinical trials and, 2, 5, 46, 56–57, 61, 64, 86, 108, 137, 154n9; discovery of, 1–2, 74, 79, 86, 98–106, 123, 140; expectations and, 61–67, 153n6, 154n9, 155n13; FDA and, 51 (see also Food and Drug Administration (FDA)); innovation and, 46, 51, 54–57; making sense of market and, 74–82, 86–87; meaning-making and, 98–106, 108–9, 120, 123; *Nature Reviews Drug Discovery* and, 74, 140; research and, 1, 5, 10, 46, 51, 55, 61–65, 74, 87, 123, 129, 132, 137, 140, 157n6; stories and, 129–30, 137, 157nn6–8; wonder, 55, 132. *See also* specific drug

Economic sociology: culture and, 6–7, 20–21; emergence and, 2, 5–6, 10, 16, 18–45, 74, 125, 127, 131–33; expectations and, 69–70, 155n13; making sense of market and, 74; meaning-making and, 99, 106–7, 108–9; markets and, 2, 18–29, 42–45, 124–127, 131–132, 151n5; stories and, 2–7, 10, 16–26, 30–34, 39–43, 69–70, 74, 100, 107, 124–27, 131–33, 151n8, 151n12, 152n15, 157n4; uncertainty and, 2, 5, 18–19, 22, 39–40, 125–26

Ehrlich, Paul, 153n7

Eli Lilly, 54–55, 158n14

Emergence: categories and, 4–5, 14–15, 19, 37–38, 41–45, 74–75, 84, 87, 89, 109–10, 121–23, 128–29, 135; cognition and, 2,

5–6, 10–13, 16–19, 25–29, 37–38, 42, 45, 84, 124–25, 128, 132–34; dynamics of, 19–29; evaluation and, 2–3, 16, 19, 26–29, 37, 40–45, 71, 74, 102, 125–28, 132, 134; expectations and, 60, 69–70, 127–29; innovation and, 2, 8–19, 27, 42, 45, 54, 60, 75, 89, 124, 131–35; making sense of market and, 74–75, 84; of markets, 2, 5–6, 10, 16, 18–45, 74, 125, 127, 131–33; market processes and, 2–6, 12, 15–31, 36–40, 43–45, 58, 60, 69–71, 74, 84, 89, 91, 123–28, 132–35; meaning-making and, 89, 94, 102, 109–10; movement in, 110–23; of newness, 3, 10–13, 16, 26–27, 29, 42, 131–32; organizational, 5, 8–12, 16, 26–27, 36, 125, 127, 134; Padgett/Powell model and, 6, 10, 26–29, 125; phases of, 3–6, 16, 31, 69–70, 84, 128–29, 133; profit and, 2, 18, 29, 37, 69, 123–24, 135; semantic analyses and, 110–23; social formations and, 19–31, 36–37, 43–44; sociocognitive actors and, 19, 25, 27, 37, 51, 55; sociology and, 6, 9–12, 17–19, 26–27, 42, 124, 127, 131, 134–35; sociomaterial actors and, 19, 25, 27, 37, 44; stories and, 18–19, 25–31, 36, 42–45, 124–35; therapeutic markets, 69–72; White and, 18, 21–30, 33, 44–45

EntreMed, 155n13

Entrepreneurship: biotechnology and, 10, 53, 140, 158n2; innovation and, 8–10, 28–29, 40, 53, 140, 158n2; *Nature's Bioentrepreneur* and, 140, 158n2; stories and, 28–29, 40

Epidermal growth factor (EGF): innovation and, 55; making sense of market and, 79–83; meaning-making and, 97

Equivalence, 25, 29, 43, 150n9

Erythropoietin (EPO), 56, 76

Estrogen, 51, 77, 83, 113, 136

European Medicines Agency (EMA), 153n6

Evaluation: emergence and, 2–3, 16, 19, 26–29, 37, 40–45, 71, 74, 102, 125–28, 132, 134; expectations and, 71–72; making sense of market and, 73–78, 85–87, 156n5; meaning-making and, 92–93, 100, 102, 106–7, 108–9, 122; new markets and, 74–83; stories and, 3–4, 16–22, 25–29, 33, 37–45, 71, 74, 87, 106–7, 125–34, 139, 151n8; therapies and, 54, 60, 62, 64

Evaluative frameworks, 28, 40–41

Expectations: about the future, 37, 38–39, 62, 68–69, 71–72, 74–75, 84, 87, 102, 123, 125–28; ambiguity and, 71; angiogenesis and, 59, 70; antibodies and, 59–63, 67; antigens and, 59–62; Batimastat and, 63–67; biotechnology and, 58–72, 153n1, 154n3, 154nn7–8, 154n10, 155n12; blockbuster drugs and, 63–65; British Biotech and, 154n10; cells and, 59–63, 67, 70, 155n11; chemotherapy and, 62–63, 66–69, 153n7, 154n9; clinical trials and, 5, 46, 56–57, 61–71, 86, 108, 137, 154n9, 155n14; cognition and, 59, 70–71; collaboration and, 17, 59–60, 71; communication and, 59–60, 71; competition and, 58, 60, 65, 69–71, 153n10; drugs and, 61–67, 153n6, 154n9, 155n13; economic sociology and, 39–40, 69–72, 125–28, 155n13; emergence and, 60, 69–72, 127–29; evaluation and, 71–72; extracellular environment and, 59, 63; findings and, 61–69; Food and Drug Administration (FDA) and, 62, 68, 153n6, 154n9, 155n11; forward-looking statements and, 104, *105*, 152n14, 156n8; Genentech and, 153nn9–10, 154n9, 155n11; genetics and, 59–60, 67–70, 154n3, 155n13; HER2 protein and, 58, 61–63, 66, 154n9; hormones and, 60; identity and, 69; immunotherapy and, 59, 62–63, 67, 70; industry analysts and, 60–62, 66, 68; inhibitors and, 59, 63, 65–66, 70, 155n12; innovation and, 58–60, 67–72; intracellular environment and, 59–60, 158n15; investors and, 62–65, 68, 71–72;

Expectations (*continued*)
journalists and, 71; language and, 133, 152n17; Marimastat and, 64–67; market analysts and, 74–83; matrix metalloproteinase (MMP) and, 59, 63, 65, 70; molecular engineering and, 60–61, 63, 71–72, 154n9; narratives and, 58, 69–71; oncology and, 59, 68; pharmaceutical industry and, 61–62, 65–66, 68, 154n8; pipelines and, 65–66, 68; positionings and, 61–69; profit and, 17, 37, 59, 62, 65, 69–72, 123, 154n10; projectivity and, 19, 37–38; proteins and, 59–60, 63, 67; regulation and, 66, 153n6; research strategies and, 59–60, 70; scientists and, 66; stories and, 58–60, 69–72; toxicity and, 62, 66; *trastuzumab* (Herceptin) and, 61, 67–69, 72, 128, 130, 156n4; tumors and, 59–63, 66–67, 70, 154n9, 155n13; uncertainty and, 62, 70–71

Extracellular environment, 59, 63

Financial Times newspaper, 106, 141, 159n3

Food and Drug Administration (FDA): diagnostics and, 86–87, 154n9; expectations and, 62, 68, 153n6, 154n9, 155n11; Genentech and, 54–55, 68, 83, *108*, 154n9, 155n11, 157n6, 157n8; Herceptin and, 68, 73, 82, 86, 88, 133; innovation and, 51, 54–56; making sense of market and, 73, 75, 82–83, 86–88, 157nn6–10, 158nn11–21; meaning-making and, 97–99, *101*; stories and, 157nn6–10, 158nn11–21

Formal analyses: CorText and, 146–48; networks and, 146–48; qualitative analyses and, 7, 15, 143; Sankey diagrams, 148; semantic network analyses and, 146–48; topic models and, 145–46

Forward-looking statements, 104, *105*, 145, 152n14, 156n8

Fourcade, Marion, 4, 19, 25

Gefitinib, 81

Genentech: Bazell on, 154n9; expectations and, 153nn9–10, 154n9, 155n11; FDA and, 54–55, 68, 83, *108*, 154n9, 155n11, 157n6, 157n8; innovation and, 53–57, 60–63, 66–70; making sense of market and, 77, 79, 81, 83; meaning-making and, 97, 107, *108*; *pertuzumab* and, 157n6; scientists and, 55, 66

Genetics: BRCA, 50, 82, 136–37, 158n17; categories and, 6; DNA, 52–55, 147n5; expectations and, 59–60, 67–68, 70, 154n3, 155n13; innovation and, 46, 50, 52–55, 153n12; making sense of market and, 77–85; meaning-making and, 90, 97–98, *105*, 113, 116, 121; mRNA, 135; mutations and, 1 (*see also* mutations); oncogene, 1, 52–53, 55, 59, 61, 153n8; pharmaceutical industry and, 53, 55, 157n2; stories and, 130, 136–38; therapies and, 46, 50, 52–55, 55, 153n12

Gibbs sampling, 94, 97, 146, 156n3

GlaxoSmithKline, 82

Gleevec, 58, 82, 87, 155n3, 156n4

HER2 protein: chemotherapy and, 63, 77, 83, 116, 119, 136, 154n9; expectations and, 58, 61–63, 66, 154n9; future research and, 136–37; hormones and, 55, 58, 61–63, 66, 73, 77–85, 97, *101*, 116, 119, 121, 136, 153n12, 154n9, 157nn6–7, 158n11, 158n16; innovation and, 55, 58, 136, 153n12; making sense of market and, 73, 77–85; matrix metalloproteinase (MMP) inhibitors and, 97; meaning-making and, 97, *101*, 116, 119, 121; mTOR protein and, 97, 116; mutations and, 81–82, 97, 136, 158n16; neu oncogene and, 55, 73, 81, *101*, 153n12; stories and, 136; therapies and, 55, 58, 63, 77, 83, 116, 118, 136, 153n12, 154n9

Her-2: The Making of Herceptin (Bazell), 58

Herceptin: expectations and, 61, 67–69, 72, 128, 130, 156n4; FDA and, 68, 73, 82, 86, 88, 133; making sense of market and, 73–88; meaning-making and, 97, 100, *101*, 107, *108*, 108, 113–14, 116; reduced commercial potential of,

77; targeted therapies and, 73, 79, 82, 85–88, 100, 108, 128, 130, 136
Hoffman-La Roche: *atezolizumab* and, 158n21; innovation and, 55–57, 68–69; making sense of market and, 76–77, 79, 81, 83; meaning-making and, 97, 107, 108; *pertuzumab* and, 157n6; *trastuzumab* and, 157n4, 157n8
Hormones: angiostatin, 155n13; estrogen, 51, 77, 83, 113, 136; expectations and, 60; HER2 protein and, 55, 58, 61–63, 66, 73, 77–85, 97, *101*, 116, 119, 121, 136, 153n12, 154n9, 157nn6–7, 158n11, 158n16; human growth (hGH), 55–56, 153n9; making sense of market and, 75–78, 81; meaning-making and, 97, *108*, 113, 116; progesterone, 136; receptors and, 6, 51, 97, 113, 136–37, 158n16; stories and, 136–37; therapies and, 1, 6, 47, 51, 55, 60, 75–78, 81, 97, 113, 116, 136–37
Human growth hormone (hGH), 55–56, 153n9
Human Tumor Atlas Network, 138

Identity: collaboration and, 13; expectations and, 69; formation of, 36–37; *Identity and Control*, 23, 131; market niches and, 150n1; meaning-making and, 122; stories and, 4, 12, 23, 26, 29–30, 36–38, 41, 44, 69, 131, 133; switching, 13; White and, 12–13, 23, 26, 29–30
Identity and Control (White), 23, 131
Imatinib, 83, 155n3
Immunotherapy: *atezolizumab* and, 158n21; expectations and, 59, 62–63, 67, 70; making sense of market and, 76–77, 82; meaning-making and, 97, *101*, 109, 116, 119, 123; PDL1, 158n21; research strategies and, 70, 77, 109; targeted therapies and, 137; tyrosine kinase receptors and, 61–62, 82, 97, 155n3
Industry analysts: categories and, 89, 96, 98–102, 109, 123; expectations and, 60–62, 66, 68; making sense of market and, 82; meaning-making and, 89, 96, 98–102, 109, 123; stories and, 3, 15, 133, 137; topic models and, 98–101
Inhibitors: CDK4/6, 137; cells and, 51, 59, 63, 80, 82, 97, *101*, *105*, 119, 137, 157n5, 158n12; expectations and, 59, 63, 65–66, 70, 155n12; kinase, 82, 103–4, *103*, *105*, 137, 155n3; making sense of market and, 77–83; matrix metalloproteinase (MMP), 59, 63, 65, 70, 77, 97, 100, *101*, 102, *103*, *105*, 109, 129, 157n11; meaning-making and, 97, 100, *101*, 102, 103, 104, *105*, 109, 119, 155n3; PDL1, 158n21; PI3K, 137; stories and, 129, 137, 157n5, 158nn12–14, 158nn18–19, 158n21; therapies and, 51
Innovation: actors and, 57; biotechnology and, 1–2, 4, 7–11, 15–17, 46–47, 53–60, 67, 126, 128, 133, 140, 153n9; British Biotech and, 53, 63–68, 70; chemotherapy and, 46, 49–51, 55, 136–37, 157nn8–9, 157n12; cognitive complexity and, 7, 10–11; cognition and, 2, 4, 7, 10–13, 19–22, 25–30, 33–34, 37–45, 124, 128, 133; collaboration and, 53–55; competition and, 53–54; diagnostic, 47, 49; digital, 135; drugs and, 46, 51, 54–57; emergence and, 2, 8–19, 27, 42, 45, 54, 60, 75, 89, 124, 131–35; entrepreneurship and, 8–10, 28–29, 40, 53, 140, 158n2; epidermal growth factor (EGF) and, 55; evolution of technology and, 8–9; expectations and, 58–60, 67–72; Food and Drug Administration (FDA) and, 51, 54–56; future research and, 134–38; Genentech and, 53–57, 60–63, 66–70; genetics and, 46, 50, 52, 52–55, 153n12; HER2 protein and, 55, 58, 136, 153n12; historical perspective on, 8; Hoffman-La Roche and, 55–57, 68–69; individual, 15; investors and, 46, 53–57; late 1980s treatments and, 53–57; "magic bullet", 46, 52–53, 56, 70, 85, 153n7; making sense of market and,

Innovation (*continued*)
75–80, 84–86; meaning-making and, 97, 108, 116; metaverse and, 135; molecular engineering and, 1, 5–6, 8, 15–16, 47, 52–53, 129, 136; oncogene and, 1, 52–53, 55, 59, 61, 153n8; Padgett and, 53; pharmaceutical industry and, 53–57; pipelines and, 20, 65–66, 68, 75, 81, 86, 99, *101*, *103*, 104, 106, 123; positionings and, 61–69; Powell and, 56–57; processual study of, 131–34; profit and, 2, 9, 37, 53, 75, 108, 124; proteins and, 54–56; radiation and, 47, 50–52; scientists and, 55, 152n3; sociology of, 7–15, 18–19, 38, 58, 131–35; Stark and, 27–29; stories and, 18–19, 24, 33–34, 37–45, 126, 128, 131–36; targeted therapies and, 6, 15, 75, 84–86, 89, 126, 128, 133, 136

Insulin, 54–55, *108*, 153n10
Interferon, 5, 46, 55–56, 78
Interleukins, 56, 97, 109, 153n15
Intracellular environment, 59–60, 158n15
Investors: biotechnology and, 1–2, 46, 53–57, 63–64, 71–72, 74, 98, 141, 158n1; expectations and, 62–65, 68, 71–72; innovation and, 46, 53–57; making sense of market and, 74, 84; meaning-making and, 98; stories and, 125–26, 132; therapies and, 46, 53–57

Iressa, 81

Journalists: categories and, 89–90, 93–94, 106–7, 108, 110, 123; expectations and, 71; meaning-making and, 89–90, 93–94, 106–7, 108, 110, 123; methodology and, 139–40, *141*, 158n1; stories and, 3–4, 8, 15, 37, 126–29, 133

Kaplan, Sarah, 40, 56, 91
Kinase inhibitors, 82, *101*, 103, *103*, *105*, 137, 155n3

Language: as data, 35; expectations and, 133, 152n17; grammar and, 93; meaning-making and, 91, 110; narratives and, 30 (*see also* narratives); natural language processing (NLP) and, 91, 111–12, 145–47, 156n11; stories and, 13, 30–34, 43, 133, 152n17; topic models and, 91

Lapatinib, 82–83, *97*, 133, 136
Latent Dirichlet Allocation (LDA): Gibbs sampling and, 94, 97, 146, 156n3; limitation of, 93; scientific discussions and, 93–97; topic models and, 91–97, 146, 156n3

Latour, Bruno, 13–14, 28, 45
Leukemias, 47, 56, 155n3, 156n4
LexisNexis Academic Universe, 140, 145
Louvain community detection algorithm, 111–12, 118–19, 147
Lung cancer, 136; categories and, 113; clinical trials and, 64; expectations and, 56, 63–64, 67; inhibitors and, 63, 82; Proleukin and, 56; topic models and, 94, *105*, *108*

Mab Thera, 82, 155n11, 156n4
"Magic bullet" treatment, 46, 52–53, 56, 70, 85, 153n7
Magic Cancer Bullet (Vasalla and Slater), 58
Making sense of market, 17; actors and, 74, 84, 87; angiogenesis and, 77–83; antibodies and, 76–86; antigens and, 79, 86; biotechnology and, 73–76, 80–82, 85–87; blockbuster drugs and, 75, 77, 80–81, 86–87; blood and, 82–83; categories and, 73–80, 84–88; cells and, 79–82, 85–86, 155n3; chemotherapy and, 75–77, 80–85, 156n6; clinical trials and, 77, 82, 86, 155n2; cognition and, 84; collaboration and, 73, 81, 87, 156n5; communication and, 84; competition and, 80; drugs and, 74–82, 86–87; emergence and, 74–75, 84; epidermal growth factor (EGF) and, 81–83; evaluation and, 73–87, 156n5; Food and Drug Administration (FDA)

and, 73, 75, 82–83, 86–88, 157nn6–10, 158nn11–21; Genentech and, 77, 79, 81, 83; genetics and, 77–85; HER2 protein and, 73, 77–85; Herceptin and, 73–88; Hoffman-La Roche and, 76–77, 79, 81, 83; hormones and, 75–78, 81; immunotherapy and, 76–77, 82; industry analysts and, 82; inhibitors and, 77–83; innovation and, 75–80, 84–86; interpreting new markets and, 74–83; investors and, 74, 84; matrix metalloproteinase (MMP) and, 77; meaning-making and, 85; models and, 80–81, 86–88; molecular engineering and, 73, 75, 79–81, 85–88; mutations and, 81–82; oncology and, 75–81, 86, 88, 156n6; patient potential and, 74–75; pharmaceutical industry and, 156nn5–6; pipelines and, 75, 81, 86; profiling and, 81–82; profit and, 75, 80, 82; proteins and, 77, 79, 83; regulation and, 6, 79, 81; stories and, 74, 84, 87–88; targeted therapies and, 73, 75, 81–88, 126–33, 136–37, 157n3, 157n5; toxicity and, 73–80, 85, 87; tumors and, 78–85, 88; uncertainty and, 76, 86

Malignant cells, 47, 80, 86, *101*

Marimastat: British Biotech and, 64–67, 100, 102–9; expectations and, 64–67; market share of, 64

Markets: analytic model of, 124–31, 135; categories and, 4–6, 123, 132–33 (*see also* categories); emergent processes and, 2–6, 12, 15–31, 36–40, 43–45, 58, 60, 69–71, 74, 84, 89, 91, 123–28, 132–35; evaluating new, 74–83; of expectations, 59, 71–72, 127; interpreting, 74–83; marketization, 24, 45; niche, 21–22, 70, 150n1; pipelines and, 20, 65–66, 68, 75, 81, 86, 99, *101*, *103*, 104, 106, 123; positionings and, 61–69; profit and, 123–24 (*see also* profit); as social formations, 19–29; stock prices, 56, 62, 64–70, 90, 99, *101*, 107, 123, 154n10, 155n13; stories and, 18 (*see also* stories)

Market analysts: 74–75; qualitative study and, 77, 84

Marx, Karl, 149n4

Mastectomy, 49–50

Matrix metalloproteinase (MMP) inhibitors: angiogenesis and, 59, 70, 77; biotechnology and, 63, 65, 70, 77; expectations and, 59, 63, 65, 70; HER2 protein and, 97; making sense of market and, 77; meaning-making and, 97, 100, *101–4*, 109; stories and, 129, 157n1; targeted therapies and, 100

McCloskey, Deirdre, 32

Meaning-making: actors and, 89, 97, 123; angiogenesis and, *105*, 116; antibodies and, 100, *101*, *105*, 107, *108*, 116, 123; antigens and, 97, *101*, 119, 123; Batimastat and, 100, *105*, 106–10; biotechnology and, 90, 96, 98, *101*, 102, 104, 106, 108, 123; British Biotech and, 102, *105*, *108*, 123; categories and, 89–123, 156n1; cells and, 90, 96–97, *101*, *105*, 110, 113, 117, 119–22, 157n12; chemotherapy and, 95, 96, 102, 104–10, 113–16, 119, 122; clinical trials and, 94, 97, *105*, 106, *108*; clusters and, 90–94, 97–100, 110–13, 116–22; collaboration and, 89–90, *105*, 108, 123; collective, 3, 13, 34, 36; communication and, 98; drugs and, 98–106, 108–9, *120*, 123; economics and, *100*, 106–7, 108–9; emergence and, 89, 94, 102, 109–10; epidermal growth factor (EGF) and, 97, 116; Food and Drug Administration (FDA) and, 97–99, *101*; formal analysis and, 90–96, 110, 111–12, 118–19, 147; Genentech and, 97, 107, *108*; genetics and, 90, 97–98, *105*, 113, 116, 121; HER2 protein and, 97, *101*, 116, 119, 121; Herceptin and, 97, 100, *101*, *107*, 108–9, 113–14, 116; Hoffman-La Roche and, 97, *107*, 108; hormones and, 97, *108*, 113, 116; identity and, 122; immunotherapy and, 97, *101*, 109, 116, 119, 123; individual, 3, 11, 15, 33; industry analysts and, 89, 96, 98–102, 109, 123; inhibitors

Meaning-making (*continued*)
and, 97, 100, *101*, 102, 104, *103*, *105*, 109, 119, 155n3; innovation and, 97, 108, 116; investors and, 98; journalists and, 89–90, 93–94, 106–7, 108, 110, 123; language and, 91, 110; Latent Dirichlet Allocation (LDA), 91–97, 146, 156n3; Louvain community detection algorithm and, 111–12, 118–19, 147; making sense of market and, 85; matrix metalloproteinase (MMP) and, 97, 100, *101*–*4*, 109; models and, 89–98, 100, 104, 109–10, 116, 119–20, 122–23, 156n1, 156nn3–4; molecular engineering and, 96, *101*, *105*, *108*, 109, 113, 116, 119–22; mTOR protein and, 97, 109, 116, 119–22; mutations and, 97; networks and, 89–91, 109–10, 113–23, 156nn10–11, 157n12; newspaper articles and, 106–10; oncology and, 17, 89, 93, 99, *100*, *101*, 104, 109–20, 156n5; over time, 91–110; pharmaceutical industry and, *101*, 104, *105*, *108*; pipelines and, 99, *101*, *103*, 104, 106, 123; profit and, 92, *108*, 123; proteins and, *101*, *105*, *108*, 113; radiation and, 111, 116; regulation and, 96, *101*, *105*, *108*; scientists and, 93–97, *108*; semantic network analyses and, 89–90, 110–23; sensemaking and, 33–34, 37, 123; stories and, 3–7, 11–15, 29–30, 33–36, 41–44, 89–91, 96, 106–7, 123, 126–27, 132–34; targeted therapies and, 89–90, 95–97, 100, *101*, 109–10, 116–23, 157n12; topic model analyses and, 91–110; toxicity and, 94–95, *96*; tumors and, 100, *101*, *105*, *108*, 110, 119–22, 156n2, 157n12; wire reports and, 102–6
Metastasis, 47, *101*, 137
Metaverse, 135
Methodology: categories and, 14, 90, 110, 133, 135, 140–42; data sources and, 139–40, 158n2, 159n3; evaluation and, 139–43; formal analyses, 145–48; industry analysts and, 141; innovation and, 7, 140; investors and, 141; journalists and, 139–40, *141*, 158n1; meaning-making and, 142; networks and, 139–43; qualitative analysis, 58, 60, 75; research design and, 139–43; semantic network analyses and, 142; text as data, 91, 134, 139–40; topic modeling, 7, 15, 17, 89–110, 116, 119–20, 122–23, 129, 134, 142, 145–46, 156n1, 156nn3–4
Mische, Ann, 3, 11, 38-39
Models: analytic, of the book, 7, 9, 16–20, 25–26, 29, 31, 34–35, 42, 91–110, 122–31, 135, 142, 145–46; cognitive, 2, 16, 19–20, 25–33, 43–44, 125, 131–34; making sense of market and, 80–81, 86–88; meaning-making and, 89–99, 104, 109–10, 116, 119–20, 122–23, 156n1, 156nn3–4; stories and, 15–36, 39, 42–45, 89–90, 123–38; theoretical, 2, 16, 31, 131; White, 21–26, 44, 131, 150nn2–4, 151n7
Mohr, John, 7, 14, 35–36, 91, 134
Molecular engineering: expectations and, 60–61, 63, 71–72, 154n9; innovation and, 1, 5–6, 8, 15–16, 47, 52–53, 129, 136; making sense of market and, 73, 75, 79–81, 85–88; meaning-making and, 96, 99, *105*, *108*, 109, 113, 116, 119–22; methodology and, 158n2
Mortality rates, 1, 47, *48*, 51, 57
mRNA, 135
mTOR protein: biotechnology and, 158n15; HER2 protein and, 97, 116; meaning-making and, 97, 109, 116, 119–22; stories and, 137
Murray, Fiona, 5, 54, 56
Mutations: BRCA, 50, 82, 136–37, 158n17; categories and, 6; expectations and, 154n3; HER2 proteins and, 81–82, 97, 136, 158n16; inherited, 47; making sense of market and, 81–82; meaning-making and, 97; oncogene, 1, 52–53, 55, 59, 61, 153n8; PARP inhibitors and, 137, 157n5, 158n17; therapies and, 47, 50, 52

Narratives: as data, 35; embedding, 33, 35; expectations and, 58, 69–71; institutionalism and, 151n13; methodology and, 143; new, 151n13; nodes and, 14, 36; projectivity and, 19, 37–38; stories and, 2–5, 14, 19, 29–40, 43, 45, 125, 127, 132–35, 150n7
National Breast Cancer Coalition, 154n9
Natural language processing (NLP), 91, 111–12, 145–47, 156n11
Nature journal, 59
Nature Reviews Drug Discovery (Business Insights), 74, 140
Network domains (netdoms), 12–13, 15, 26–28, 43-45, 124, 129
Networks: actor-network theory (ANT), 11–15, 34, 150nn8–9, 151n5; analysis of, 3, 6–8, 11–20, 25–30, 34, 36, 44–45, 60, 89–91, 109–10, 116, 121–29, 142–43, 146–47, 149n4, 150n3, 150n9, 156n11; cognition and, 2, 4, 10–13, 16, 19–22, 25, 27, 37, 44, 124, 128, 130–33; collaboration and, 3, 13, 17, 60, 89–90, 123, 128–29, 133, 138, 150n9; embedded, 20, 23, 91; expectations and, 60; folding, 10, 12, 27, 129, 135; formal analyses and, 146–48; interdependence and, 4, 30, 131, 149n4; as lenses, 21; Louvain community detection algorithm and, 111–12, 118–19, 147; market perspective and, 2–3; meaning-making and, 89–91, 109–10, 113–23, 156nn10–11, 157n12; methodology and, 139–43; nodes and, 112–13, 116–21, 147–48, 150n9, 157n12; relational sociology and, 7–15, 19–20, 127, 131, 134, 142, 149n4, 150n1; semantic, 7, 15, 17, 89–91, 110–23, 129, 134, 142, 146–48, 157n12; stories and, 3–4, 11–12, 15–30, 34–37, 43–45, 60, 89–90, 123–39, 150n1, 150n3
Neu oncogene, 55, 73, 81, *101*, 153n12
New Cancer Therapeutics (Dooley), 78
New York Times, 106, 141
Niches, 21–22, 150n1

Nodes: natural language processing (NLP) and, 111–12; networks and, 112–13, 116–21, 147–48, 150n9, 157n12; semantic network analyses and, 112–13, 116–21
Non-Hodgkin's lymphoma, 155n11, 156n4
Noninvasive treatments: effectiveness of, 149n1; genetics and, 136; "magic bullet", 46, 52–53, 56, 70, 85, 153n7; molecular engineering and, 5; target cells and, 46, 52, 55, 59, 61, 79–82, 85–86, 90, 96, *101*, *105*, 113, 117, 119–22, 137, 148, 157n12, 157n5; toxicity and, 1, 87
Novartis, 76, 155n3, 158n13, 158nn15–16

Oncogene, 1, 52–53, 55, 59, 61, 153n8
Oncology: expectations and, 59, 68; making sense of market and, 75–81, 86, 88, 156n6; meaning-making and, 17, 89, 93, 99, *100*, *101*, 104, 109–20, 156n5; methodology and, 139–42, 148; stories and, 129, 157n4; therapies and, 16, 47
Ovarian cancer, 50, 56, 63–67, 94, 136

Padgett, John: emergence and, 6, 10, 26, 29, 125; innovation and, 56–57; Powell and, 6, 10, 26, 29, 125
PARP inhibitors, 137, 157n5, 158nn17–18
Patents, 8, 106, 130, 157n4
Patient activism, 152n1
Personalized medicine, 75, 85, 87, 130
Pharmaceutical industry: clinical trials and, 2 (*see also* clinical trials); competition and, 53–54, 65; drugs, 1–2 (*see also* drugs); expectations and, 61–62, 65–66, 68, 154n8; genetics and, 53, 55, 157n2; innovation and, 53–57; making sense of market and, 75–76, 80–82, 86–87, 156nn5–6; meaning-making and, *101*, 104, *105*, *108*; patents and, 8, 106, 130, 157n4; positionings and, 61–69; spillover in, 10; stories and, 129–30, 133, 157n2. *See also* specific company
Pharmacogenetics, 157n2
PI3K inhibitor, 137

Pipelines: expectations and, 65–66, 68; markets and, 20, 65–66, 68, 75, 81, 86, 99, *101*, *103*, 104, 106, 123; meaning-making and, 99, *101*, *103*, 104, 106, 123

Powell, Walter: 8, 11, 15, 20, 151n13; emergence and, 6, 10, 26, 29, 53-54, 125; innovation and, 54, 56–57; Padgett and, 6, 10, 26, 29, 125

Product approval, 5, 116, 133, 136

Profiling, 81–82

Profit: emergence and, 2, 18, 29, 37, 69, 123–24, 135; expectations and, 17, 37, 59, 62, 65, 69–72, 123, 154n10; innovation and, 2, 9, 37, 53, 75, 108, 124; making sense of market and, 75, 80, 82; meaning-making and, 92, *108*, 123; stories and, 18, 21, 29, 37, 124, 132, 135; therapies and, 53–54

Projectivity, 19, 37–38

Proleukin, 56

Promissory organizations, 37, 74

Proteins: blocking, 59–60, 63, 137, 158n15; EGF, 55 (*see also* epidermal growth factor (EGF)); expectations and, 59–60, 63, 67; HER2, 63 (*see also* HER2 protein); innovation and, 54–56; making sense of market and, 77, 79, 83; meaning-making and, *101*, *105*, *108*, 113; stories and, 137, 157n6, 158n15; therapies and, 54–56

Radiation: invasive nature of, 1, 46, 50–51, 57; meaning-making and, 110, 113, 116

Regulation: clinical trials and, 2 (*see also* clinical trials); expectations and, 66, 153n6; making sense of market and, 6, 79, 81; meaning-making and, 96, *101*, *105*, *108*; product approval and, 5, 116, 133, 136; stories and, 126, 137; therapies and, 1–2, 5, 56

Relational sociology, 11–15, 58, 126, 131, 149n4

Rituxan, 77, 82, *108*, 155n11, 156n4

Sankey diagrams, 121, 148

Schering-Plough, 55i

Science journal, 59

Scientists: categories and, *108*; expectations and, 66; Genentech and, 55, 66; increasing knowledge of, 49; innovation and, 55, 152n3; meaning-making and, *108*; methodology and, 138; stories and, 3, 8, 14, 128; targeted therapies and, 86, 128, 138; topic models and, 93–97

Semantic network analyses: categories and, 89–90, 110–23; clusters and, 110–13, 116–22; emergence and, 110–23; formal analyses of, 146–48; groups and, 110–13; Latent Dirichlet Allocation (LDA) and, 93; Louvain community detection algorithm and, 111–12, 118–19, 147; meaning-making and, 89–90, 110–23; methodology and, 142; natural language processing (NLP) and, 111–12, 147; networks and, 7, 110–23; nodes and, 112–13, 116–21; research strategies and, 112–16; stories and, 7, 15, 17, 129, 134; targeted therapies and, 110–11, 116–23; topic models and, 92–93

Sensemaking, 33–34, 37, 123

Simmel, Georg, 110, 149n4

Social formations: emergence and, 19–31, 36–37, 43–44; expectations and, 71; markets as, 19–29; stories and, 4, 6, 12–31, 36–37, 43–44, 124–32

Sociocognitive actors: emergence and, 19, 25, 27, 37, 51, 55; stories and, 2, 6, 12–13, 16, 19, 25, 27, 37, 51, 55, 124, 128, 130, 132

Sociomaterial actors: emergence and, 19, 25, 27, 37, 44; stories and, 2, 6, 12–13, 16, 19, 25, 27, 37, 44, 124, 130, 132

Stanford Natural Language Processing Group, 146

Stanford Topic Modeling Toolkit (STMT), 156n3

Stark, David: reflexive cognition, 27–29, 125; cognitive interdependence, 29

Statistics, 74, 92–93, 149n4, 152n2, 155n14

Status, 21, 51, 86, 113, 125, 133
Stories: analytic model and, 3–6, 16, 19, 29, 36–37, 43–45, 74, 89, 124–25, 127, 130, 132, 135; ambiguity and, 5, 19, 28, 34, 37, 40–41, 45, 71, 128, 152n18; biotechnology and, 2–4, 15, 58, 60, 69–71, 74, 87, 96, 123, 126–30, 133, 139, 153n9; categories and, 3–5, 17, 19, 28–45, 74, 84, 87–90, 123, 126–36; clinical trials and, 129, 137, 157n5; clusters of agreement and, 41; cognition and, 4–6, 16, 19, 26, 30, 33, 37–41, 44–45, 71, 84, 124, 128–33; collaboration and, 23–24, 128–29, 132–33, 138, 151n8; communication and, 3, 13, 16, 30–38, 42–43, 125, 127, 149n4; competition and, 4–5, 16, 21–24, 32, 40, 42, 44–45, 58, 60, 69–71, 125–28, 132, 135, 150n1, 151n8; context and, 4–5, 12, 15, 29, 33, 38–39, 44, 69, 91, 126, 132–33; control and, 23, 29–30, 34, 131, 158n15; as devices and data, 5–6, 16, 19, 29–31, 33–36, 37, 39, 43, 60, 74, 90, 125, 131; drugs and, 129–30, 137, 157nn6–8; economic sociology and, 2–7, 10, 16–26, 30, 32, 34, 39–43, 69–70, 74, 124–27, 131–33, 151n8, 151n12, 152n15, 157n4; economics and, 32, 100, 106–7; emergence and, 2–6, 12, 15–31, 36–45, 58, 60, 69–71, 74, 84, 89, 91, 123–35; entrepreneurship and, 28–29, 40; evaluation and, 3–4, 16–22, 25–29, 33, 37–45, 71, 74, 87, 106–7, 125–34, 139, 151n8; expectations and, 58–60, 69–72; framework of markets and, 7, 11, 16, 26, 29; of the future, 2–5, 18–19, 33, 37–45, 70–74, 84, 87–88, 123–30, 134–38, 152n16; genetics and, 130, 136–38; hormones and, 136–37; identity and, 4, 12, 23, 26, 29–30, 36–38, 41, 44, 69, 131, 133; industry analysts and, 3, 15, 133, 137; inhibitors and, 129, 137, 157n5, 158nn12–14, 158nn18–19, 158n21; innovation and, 18–19, 24, 33–34, 37–45, 126, 128, 131–36; interpretation of, 3–4, 12, 15–16, 29, 33, 36–45, 60, 71, 74, 87, 106, 125–32, 138; investors and, 125–26, 132; journalists and, 3–4, 8, 15, 37, 126–29, 133; language and, 13, 30–34, 43, 133, 152n17; making sense of market and, 74, 84, 87–88; markets and, 4, 6, 12–31, 36–37, 43–44, 124–32; matrix metalloproteinase (MMP) and, 129, 157n1; meaning-making and, 3–7, 11–15, 29–30, 33–36, 41–44, 89–91, 96, 106–7, 123, 126–27, 132–34; models and, 19–36, 39, 42–45, 124–38; mTOR protein and, 137; narratives and, 2–5, 14, 19, 29–40, 43, 45, 125, 127, 132–35, 150n7, 151n10, 151n13; networks and, 3–4, 11–12, 15–30, 34–37, 43–45, 60, 89–90, 123–39, 150n1, 150n3; oncology and, 129, 157n4; pharmaceutical industry and, 129–30, 133, 157n2; public, 18–19, 37–38, 69, 125; profit and, 18, 21, 29, 37, 124, 132, 135; projectivity and, 19, 37–38; regulation and, 126, 137; research methodology and, 2–3, 7, 11, 15, 31, 36, 43, 45, 89, 124, 127, 133, 139; scientists and, 3, 8, 14, 128; semantic network analyses and, 7, 15, 17, 129, 134; sociocognitive actors and, 2, 6, 12–13, 16, 19, 25, 27, 37, 51, 55, 124, 128, 130, 132; sociology and, 6–7, 11, 15, 17–20, 24–27, 30–31, 38, 42, 52n16, 124, 127, 131, 134–35, 152n16; sociomaterial actors and, 2, 6, 12–13, 16, 19, 25, 27, 37, 44, 124, 130, 132; storytelling and, 30, 32; uncertainty and, 18–25, 28–29, 33, 37, 39–41, 44, 125–26, 135, 152n18; White and, 18, 21–30, 44–45, 131, 150nn1–4, 150n7, 151n7, 152n18
Sugen, 61, 66, 153n14
Surgery, 1, 49–51
Switching, 13

Targeted therapies: blockbuster drugs and, 130; categories and, 6, 15, 17, 73, 75, 80, 84–90, 95–97, 100, *101*, 109–10, 116–23, 126–33, 136, 157n12; construction of, 89–90; Herceptin and, 73, 79, 82, 85–88, 100, 108, 128, 130, 136;

Targeted therapies (*continued*)
immunotherapy and, 137; innovation and, 6, 15, 75, 84–86, 89, 126, 128, 133, 136; institutionalization of, 90; making sense of market and, 73, 75, 81–88, 126–33, 136–37, 157n3, 157n5; matrix metalloproteinase (MMP) and, 100; meaning-making and, 89–90, 95–97, 100, *101*, 109–111, 116–23, 157n12; precision/flexibility of, 89; scientists and, 86, 128, 138; semantic network analyses and, 110, 116–23; topic models and, 95–97, 100, *101*, 109–10

Text as data, 3, 16, 37, 91, 134, 139–40

Therapies: antigens and, 55; biotechnology and, 1–3, 5, 16–17, 46–47, 53–63, 67–74, 85–87, 90, *101*, *105*, 123, 127–29, 133, 139–40, 158n2; blockbuster drugs and, 6, 63–65, 75, 77, 80–81, 86–87, 129–32, 157n4; blood and, 47–49, 52, 56; categories and, 4, 6, 15, 17, 19, 73–76, 80, 84–90, 95–97, 100, *101*, 109–10, 116–23, 126–33, 136, 140, 147, 157n12; cells and, 46–56; clinical trials and, 46, 50, 55–57, 152n1; collaboration and, 53–55; competition and, 53–54; diagnostics and, 1, 47, 49, 82, 86–87, 102, 152n1, 154n9, 156n4, 157n3; drugs and, 46 (*see also* drugs); emerging market of, 69–72; evaluation and, 54, 60, 62, 64; genetics and, 46, 50–55, 153n12; HER2 protein and, 55, 58, 63, 77, 83, 116, 119, 136, 153n12, 154n9; historical perspective on, 8, 48–53; hormones and, 1, 6, 47, 51, 55, 60, 75–78, 81, 97, 113, 116, 136–37; innovation and, 46–57 (*see also* innovation); investors and, 46, 53–57; late 1980s treatments and, 53–57; "magic bullet", 46, 52–53, 56, 70, 85, 153n7; molecular engineering and, 1, 5–6, 8, 15 (*see also* molecular engineering); monoclonal antibodies (mAbs) and, 46, 55, 156n6; mutations and, 47, 50, 52; noninvasive treatments and, 1, 5, 46, 87, 136, 149n1; oncogene and, 1, 52–53, 55, 59, 61, 153n8; oncology and, 16, 47; patient activism and, 152n1; profit and, 53–54; proteins and, 54–56; regulation and, 1–2, 5, 56; research strategies and, 59–60, 70; toxicity and, 1, 46, 50–51, 55, 57; tumors and, 47–55, 130, 137–38, 153n12, 158n19. *See also* specific treatment

Theratope, 62–63, 67–69, 77, 155n2, 155n14

Thévenot, Laurent, 25, 27–28, 151nn8–9

Tobacco, 52

Topic models: 90–110; clusters and, 91–94, 97–100; computational social science and, 91–92, 134, 142; formal analyses and, 145–46; Gibbs sampling and, 94, 97, 146, 156n3; as heuristic tools, 93, 99; industry analysts and, 98–101; Latent Dirichlet Allocation (LDA), 91–97, 146, 156n3; machine learning and, 91; meaning-making and, 89–110, 116, 119–20, 122–23, 156n1, 156nn3–4; methodology and, 145–46; natural language processing and, 91, 145–46; newspaper articles and, 106–10; scientific discussions and, 93–97; semantic analyses and, 92–93; Stanford Topic Modeling Toolkit (STMT), 156n3; stories and, 7, 15, 17, 129, 134; targeted therapies and, 95–97, 100, *101*, 109; trajectories of meaning over time and, 91–110; wire reports and, 102–6

Toxicity: expectations and, 62, 66; making sense of market and, 73–80, 85, 87; meaning-making and, 94–95, *96*; therapies and, 1, 46, 50–51, 55, 57

Trading zone, 87, 122

Translation, 13–14, 28, 44, 150n8

Trastuzumab: as aging blockbuster, 157n4; biosimilars and, 157n4; empirical analyses of, 133, 136–37; expectations and, 58, 61, 67–69, 72, 128, 130, 156n4; FDA approval of, 68, 73, 82, 86, 88, 133; as groundbreaking treatment, 58; making sense of market and, 73–88; meaning-making and, *97*, *101*, 113–14,

116; *pertuzumab* and, 157n6; as target therapy, 73, 79, 82, 85–88, 100, 108, 128, 130, 136. *See also* Herceptin

Triple-negative breast cancer (TNBC), 136–37, 157n5

Tumorigenesis, 137

Tumors: expectations and, 59–63, 66–67, 70, 154n9; growth of, 15, 47, 50–52, 55, 59–61, 66–67, 82–83, *101*, *105*, 147, 153n12; Human Tumor Atlas Network and, 138; making sense of market and, 78–85, 88; malignant, 47, 80, 86, *101*; meaning-making and, 100, *101*, *105*, *108*, 110, 119–22, 156n2, 157n12; metastasis and, 47, *101*, 137; overexpressing, 78; therapies and, 47–55, 130, 137–38, 153n12, 158n19; tyrosine kinase receptors and, 61–62, 82, 97, 155n3

Tykerb/Tyverb, 82–83, 97, *97*, 133, 136

Tyrosine kinase receptors, 61–62, 82, 97, 155n3

Ullrich, Axel, 153n14

Uncertainty: ambiguity and, 5, 19, 28, 34, 37, 40–41, 45, 71, 128, 152n18; decision-making and, 2, 5, 18, 21; economic, 2, 5, 18–19, 22, 39–40, 125–26; expectations and, 62, 70–71; forward-looking and, 33, 38; making sense of market and, 76, 86; stories and, 18–25, 28–29, 33, 37, 39–41, 44, 125–26, 135

Vascular endothelial growth factor (VEGF), 79–80, 83, 97, 116

Viruses, 47

Web of Science, 93, 140, 145–46, 148

White, Harrison: competition and, 21–22; economic sociology and, 18, 21–30, 44–45, 131; emergence and, 18, 21–30, 44–45; identity and, 12–13, 23, 26, 29–30; *Identity and Control*, 23, 131; market model of, 21–26, 44, 131, 150nn2–4, 151n7; niches and, 21–22, 150n1; one-way mirror of, 22; stories and, 18, 21–30, 44–45, 131, 150nn1–4, 150n7, 151n7, 152n18; W(y)-model, 150n4

Wire reports, 102–6, 141

Wonder drugs, 55, 132

CULTURE AND ECONOMIC LIFE

Diverse sets of actors create meaning in markets: consumers and socially engaged actors from below; producers, suppliers, and distributors from above; and the gatekeepers and intermediaries that span these levels. Scholars have studied the interactions of people, objects, and technology; charted networks of innovation and diffusion among producers and consumers; and explored the categories that constrain and enable economic action. This series captures the many angles in which these phenomena have been investigated and serves as a high-profile forum for discussing the evolution, creation, and consequences of commerce and culture.

Supercorporate: Distinction and Participation in Post-Hierarchy South Korea
Michael M. Prentice
2022

Black Culture, Inc.: How Ethnic Community Support Pays for Corporate America
Patricia A. Banks
2022

The Sympathetic Consumer: Moral Critique in Capitalist Culture
Tad Skotnicki
2021

Reimagining Money: Kenya in the Digital Finance Revolution
Sibel Kusimba
2021

Black Privilege: Modern Middle-Class Blacks with Credentials and Cash to Spend
Cassi Pittman Claytor
2020

Global Borderlands: Fantasy, Violence, and Empire in Subic Bay, Philippines
Victoria Reyes
2019

The Costs of Connection: How Data is Colonizing Human Life and Appropriating It for Capitalism
Nick Couldry and Ulises A. Mejias
2019

The Moral Power of Money: Morality and Economy in the Life of the Poor
Ariel Wilkis
2018

The Work of Art: Value in Creative Careers
Alison Gerber
2017

Behind the Laughs: Community and Inequality in Comedy
Michael P. Jeffries
2017

Freedom from Work: Embracing Financial Self-Help in the United States and Argentina
Daniel Fridman
2016

Lightning Source UK Ltd.
Milton Keynes UK
UKHW020621211022
410822UK00004B/88